Sustainable Water Management

Ken'ichi Nakagami • Jumpei Kubota
Budi Indra Setiawan
Editors

Sustainable Water Management

New Perspectives, Design, and Practices

Springer

Editors
Ken'ichi Nakagami
College of Policy Science
Ritsumeikan University
Ibaraki, Osaka, Japan

Jumpei Kubota
Research Institute for Humanity
and Nature (RIHN)
Kita-ku, Kyoto, Japan

Budi Indra Setiawan
Faculty of Agricultural Technology
Bogor Agricultural University
Bogor, Indonesia

ISBN 978-981-10-9344-9 ISBN 978-981-10-1204-4 (eBook)
DOI 10.1007/978-981-10-1204-4

Printed on acid-free paper

This Springer imprint is published by Springer Nature
The registered company is Springer Science+Business Media Singapore Pte Ltd.

Preface: A New Horizon of Sustainable Water Management

The concept of integrated water resources management (IWRM) has existed for more than 70 years, but the approach was criticized by water scientists before it was completed. International organizations such as the Organization for Economic Co-operation and Development (OECD) and the World Water Council (WWC) presented new methods and additional factors related to the new direction of IWRM at the 7th World Water Forum (WWF7) in Korea. This book examines current concepts, methodology, and water management practices and introduces the new IWRM paradigm.

Three important visions will be discussed:

1. A "theory" vision that presents a new horizon for IWRM;
2. A "social implementation" vision that includes the design of local water resources management frameworks and IWRM practices (the "Designing Local Frameworks for Integrated Water Resources Management" Project of the Research Institute for Humanity and Nature, Japan);
3. A "harmony between science and society" vision that promotes a new water-recycling system featuring reclamation and reuse (the "Research Center for Water Reclamation and Recycle in Asia—For Sustainability Water Resource Development" Project of the Ritsumeikan–Global Innovation Research Organization (R-GIRO), Ritsumeikan University, Japan).

The IWRM manages surface, ground, and virtual waters as an integrated water resource to deal with the present water crisis due to climate change. Similar sustainable and comprehensive concepts and approaches should be considered to resolve various problems. Interdisciplinary approaches that integrate knowledge and information from diverse scientific fields should also be established to examine IWRM.

Continued increased demand for land and water to support numerous aspects of life has led to a growing concern regarding the availability and quality of water resources. Weather uncertainty resulting from climate change at both global and local scales forces us to evaluate what has been done in the past and rethink what should be done in the future to manage water resources that have persistently deteriorated over time. Changing land use from highly dense vegetation to more open land cover threatens the ability of rainwater to penetrate soil layers and increases soil erosion and run-off into nearby rivers and bodies of water. Rivers frequently flood downstream as a result of intensive rainfall within short periods of time. Meanwhile, base flow of the rivers has been depleted due to longer dry seasons. This situation has been exaggerated by the increased frequency of El Niño events. The most severe El Niño episode was recorded in 2015 in southern Indonesia. Not only did harvests fail in more than 200,000 paddy fields, but planting times were also delayed by 2–3 months. These events threaten staple food production. The Ministry of Agriculture leads all segments of Indonesian society including farmers, scientists, academics, and the military in an urgent effort to cultivate farmland in areas where the water supply is declared sufficient. This concerted effort illustrates the importance of stakeholder consolidation and management and the application of available methods and technologies.

As available water becomes more uncertain, scarce, and competitively sought among water users, it is important for stakeholders to gain a comprehensive understanding of the climate and seasonal changes occurring in their area and how to deal with the associated socioeconomic impacts. This book attempts to consolidate the efforts made thus far and the results of collaborative research among academics, industrial scientists, officials, and administrators from: (1) the Research Institute for Humanity and Nature, Kyoto, Japan; (2) Bogor Agricultural University, Indonesia; (3) Hasanuddin University, Makassar, Indonesia; (4) Udayana University, Denpasar, Indonesia; and (5) the National Agency for Meteorology, Climatology and Geophysics, Indonesia. This research was conducted over a 5-year period from 2011 to 2016 in the Saba and Jeneberang watersheds of Indonesia. The main objective was to produce a local framework of integrated water resources management and to focus on the identification of (1) significant climate, land use, and socioeconomic changes; (2) the interrelationship between climate, land use, and socioeconomic change related to water resources; and (3) enhancements to water-use efficiency and water productivity.

The Saba watershed is located in northern Bali, and the Jeneberang watershed is in the South Sulawesi Province of Sulawesi Island, Indonesia. The Jeneberang watershed has a large modern dam, referred to as Bili-Bili, whose irrigation system is managed according to national law. Meanwhile, the Saba watershed has the newly established Titab Dam that is smaller than the Bili-Bili and has an irrigation system managed according to the traditional law known as Subak. A preliminary study found clear indications that both Saba and Jeneberang watersheds have experienced climate and land-use changes that severely impact water users. The annual rainfall in both watersheds decreased and the dry season has been extended. A significant decrease in the water level of the Bili-Bili Dam was observed using remote sensing data. Land conversion from paddy to non-paddy fields has occurred because of an extended period of insufficient irrigation water. An effective water management strategy that meets the increasing demands of numerous socioeconomic activities as the amount of available resource decreases must be designed and involve stakeholders who can consider data from scientifically comprehensive and comparative studies.

The "harmony between science and society" vision of sustainable water management is also highly valued. This book explains that vision through the example of an ongoing research project that promotes sustainable water resource development in Asia. The Ritsumeikan–Global Innovation Research Organization (R-GIRO) of Ritsumeikan University, Japan, is the main force of this joint project that is a collaboration with the Ritsumeikan Asia Pacific University, Japan; Tongji University, China; and the East China Normal University, China. The goal of this project is to implement a new water recycling system that features reclamation and reuse. This research is currently performed on Chongming Island, China, which is administered by Shanghai city, the economic center of China. This is the world's largest alluvial island and it has plentiful water resources and unparalleled ecological value. However, Chongming followed a very dangerous trajectory in the history of its development. Neither the local nor the Shanghai municipal government has ever proposed an effective master plan. As a result, irrational actions and inappropriate behavior pose a great threat to water resources and have already drastically degraded the island's ecological value. The inherent vulnerability of this island, as well as conflicts between social and economic development and environmental conservation, must be considered. The advent of the twenty-first century marks a turning point in the history of Chongming Island. The Shanghai municipal government, in compliance with instructions from the Chinese central government to build a national ecological demonstration zone, is preparing to take action and considering all advanced approaches offered for building a Chongming "eco-island". We have provided promising pathways for regional management and environmental policy design through the introduction of comprehensive practices that promote water reclamation and reuse. Our contributions are recorded in this book as a useful reference for the relevant decision/policy makers.

This book comprises five parts. Part I describes new perspectives in the reconsideration of integrated water resources management from the point of view of design science. Part II explains water resources management designs that include the collaborative activities of multiple stakeholders: (1) future water resources management design with a participatory approach toward environmental management and assessment in collaboration with local people; (2) environmental assessment in collaboration with local people; (3) local water and soil conservation assessment based on the land-use scenario; and (4) development and application of the geographic information system (GIS) tool featuring "think together" with stakeholders. Part III presents current practices in Indonesia that use the participatory approach toward sustainable water resources management: (1) to enhance participation from multiple stakeholders in the Saba River Basin; (2) hydro-geochemical assessment of the contribution of the caldera lake to paddy irrigation and river water-level stability; (3) reconsideration of the meaning of dam construction for water resources management and environmental impact assessment of the Titab Dam project to the futurability of the Saba River Basin; and (4) channeling people, science, and water through transdisciplinary practices in South Sulawesi. Part IV describes practices on Chongming Island and Chinese regional management and environmental policies for water reclamation and recycling: (1) the current water management situation on Chongming Island; (2) characteristics of eutrophication and correlation with algae in Chongming Island's artificial river network; and (3) the impact of development on land use and the water environment. Part V provides conclusions focused on the future of sustainable water management.

This book presents the scientific foundation for integrated water resources management development and introduces research possibilities for others to collect more data and information from the two respective watersheds. Recently, the governor of Bali Province established a stakeholder forum of the Saba watershed referred to as "Forum DAS Saba" after formation through a series of stakeholder meetings. This book describes how the stakeholders formed and established their forum and how scientists facilitated and prepared scientific data and information to support their focus and decision making. This book is also a historical reference for the associated stakeholders because most of the authors and the stakeholders met through research planning, study implementation and dissemination of results through a series of on-site meetings. Continued interaction related to additional topics has been arranged, and the future sustainability of human relationships among the participants is also expected to result from this project.

We hope this book is not the final edition and that it will serve as a milestone in the advancement of science, technology, and socioeconomic aspects toward improved IWRM at the watershed scale. We welcome anyone interested in joining us on this journey and in contributing to the improvement of the information in this book.

We appreciate the local people of the Saba and Jeneberang watersheds who have shared their knowledge and their fields for environmental measurements and participated in meetings, field surveys, and other administrative work. We also thank the local governments of the Buleleng and Tabanan regencies for permission to conduct research and for all of their support and assistance.

The authors thank the Research Institute of Humanity and Nature, Kyoto, Japan, for support in research implementation and collaboration with Bogor Agricultural University (IPB), Indonesia. Special thanks are due to Dr. Andi E. Sakya, Head of the Meteorology Climatology and Geophysics Board of Indonesia; Dr. Ir. Moch. Amron, M.Sc., Head of the Indonesian Water Partnership; Np. Rahadian, Executive Director of NGO Rekonvasi Bhumi, Indonesia; the Head of the Forestry Office of Bali Province; and Edi Rachman, Operational and Commercial Director of PT Krakatau Tirta Industry, Indonesia, for their contributions to local community improvement in the Saba River Basin and increasing stakeholder ability to understand the IWRM concepts. We also appreciate the Ritsumeikan–Global Innovation Research Organization, Ritsumeikan University, for promoting the collaborative study between Japan and China. Finally, we would like to express our gratitude to Dr. Xiaochen Chen and Dr. Hisaaki Kato for their assistance and coordination roles in the entire publication process.

Ibaraki, Osaka, Japan — Ken'ichi Nakagami
Kita-ku, Kyoto, Japan — Jumpei Kubota
Bogor, Indonesia — Budi Indra Setiawan

Contents

Part I
New Perspectives

Chapter 1
New Perspectives: Reconsideration of IWRM from the Viewpoint of Design Science

Ken'ichi Nakagami

Abstract Integrated water resources management (IWRM) is a key concept of water management in the twentieth century. The definition developed by the Global Water Partnership in 2000 is too penetrating the academic and decision-maker society. Although the concept of IWRM has a long history of more than 70 years, many critiques of the currently accepted definition have been generated by water scientists. Reconsideration of IWRM from the viewpoint of design science and especially sustainability science is important for creating new perspectives of IWRM. The use of IWRM toward a sustainable society was examined in relation to water sustainability and the evaluation of a water resources development project with sustainability. During the 7th World Water Forum (WWF7) in Korea in 2014, the Organization for Economic Co-operation and Development (OECD) proposed "Principles on Water Governance: From Vision to Action," and the World Water Council (WWC) reported a discussion paper entitled "Integrated Water Resource Management: A New Way Forward." Through the review of these international trends, the largest impact factor affecting IWRM is climate change. The influence of climate change on water issues became clear by the International Panel on Climate Change Fourth Assessment Report (IPCC AR4), and a strategic adaptation plan and methods for IWRM were proposed.

Keywords Integrated water resources management (IWRM) • Design science • Water demand • Adaptive management • Sustainability • Dam • Society and human system

---

K. Nakagami (✉)
College of Policy Science, Ritsumeikan University,
2-150 Iwakura-cho, Ibaraki, Osaka 567-8570, Japan
e-mail: nakagami@sps.ritsumei.ac.jp

K. Nakagami et al. (eds.), *Sustainable Water Management*,
DOI 10.1007/978-981-10-1204-4_1

1.1 Introduction

The twenty-first century has been called the "century of water," in contrast with the twentieth century that has been referred to as the "century of oil." The term the "century of water" refers to the concept of overcoming the competition of resources generated during industrial development, rural exploration, and war. However, obtaining water resources "is a much more intensive process than obtaining oil," and the deterioration of the daily water environment is threatening human health and survival, with the prospect of "water wars" occurring in some regions in the short future.

Entering the twenty-first century, the phenomena caused by water shortages have been considered as social problems and thus have attracted the attention of researchers. The normalization of recorded heavy rainfall, high frequency of extreme disasters, and the expansion of water pollution are striking. Today, there are approximately 0.7 billion people in the world experiencing water shortages. Moreover, 4700 children annually pass away because of a lack of access to sanitary water. Who is responsible for protecting water sources in order to save lives? The United Nations (UN) has been attempting to answer this humanitarian question, but has had few significant results. Also, the globalization of economy has made it possible for drainage construction in developed industrial countries and developing countries to be operated globally as a water business. Regarding current water problems, the search for a solution from a policy science view is vital when constructing water resources environmental policies. Through this process, strategies of constructing a sustainable society as the base of water security are desirable. To resolve this difficult situation, the concept of Integrated Water Resources Management (IWRM) has performed an important role. The core value of the 6th World Water Forum (WWF6) in 2012 entitled "Time for Solutions" was discussed seriously by 33,000 participants. The IWRM and core value of the 7th World Water Forum (WWF7) in 2015 entitled "Water for Our Future" proposed the next direction of IWRM. At the same time, the UN World Water Development Reports have examined the future direction of the IWRM. Sustainable Development Goal (SDG) 6 entitled "Ensuring the availability and sustainable management of water and sanitation for all" is a new milestone for seeking a new paradigm of IWRM.

The perception of water resources and environmental policy, as well as the establishment of a sustainable society based on water security insurance, is a very important policy topic in the twenty-first century.

1.2 Water Crisis and Water Security

1.2.1 Aspects of the Contemporary Water Issues

What is a water issue? This question is the reflection of the contemporary world conflict. The framework of the contemporary water issues is illustrated in Fig. 1.1.

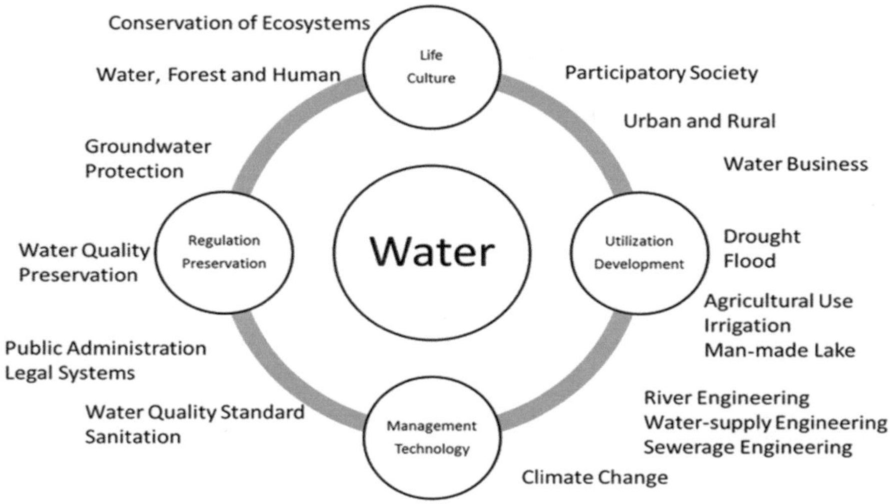

Fig. 1.1 Illustration of contemporary water issues

Water is the source of life for industry and community. To understand the water issues as a scientific field, the starting step is life and culture-related water uses. A long history of man and water has created and formulated civilizations. The famous phrase, "The Sea is longing for the forest (mori wa umi no koibito)" by Mr. Hatakeyama Shigeatsu is a symbolic idea that indicates the relationship between Japanese historical and traditional man and water. We could find the wisdom of our ancestors by his phrase.

The main objectives of the River Act, enacted in April 1896, were flood control and navigation, and the New River Act enacted in July 1964 mainly focused on water utilization. More recently, the Revised New River Act enacted in June 1997 stressed the overall maintenance of three factors, including the maintenance and preservation of river environments, reflection of regional opinions, and introduction of planning systems (ITO 2012).

To enact changes in the River Act, which had separated the local residents from the river management, the fundamental democratic question on the participation of citizens was required. After the enforcement of the Revised New River Act, drastic changes had been started on the construction and management of a dam, which had concerns by local residents. The relationship between man and water drastically changed in the twentieth century, and this corresponded to the change of the relationship between urban and rural areas.

The appropriate use and development of water means setting a baseline for urban water management and developing a special urban citizen task force to address potential water shortages and flood control. The role of traditional agriculture, irrigation, and man-made lakes has been rapidly altered after high economic growth periods in Japan. Based on these trends, river engineering, water supply engineering, and sewage engineering in related urban areas have rushed headlong into the maintenance of infrastructure. Moreover, by the progress of the managing technology, the water projects have been operated by the government using sophisticated systems and criteria. However, the effects of climate change have been transformed into a reality, and thus the limits of the traditional management system should be investigated by conducting a strategic and policy review concerning the strategic adaptation. Moreover, the maintenance and renewal of waterworks facilities will become an urgent task, and the influence of the burden of the budget of municipal governments will become serious.

An individual system has been made related to the regulatory and administrative frameworks of water, resulting in the comprehensive water law entitled "Basic Law on the Water Cycle" enacted on April 2, 2014.

The Cardinal principles are as follows:

1. Seeing that water sustains life on earth and plays an important role in the livelihoods of people and in industrial activities in the course of the sound watery rotation, it must be encouraged positively to tackle its maintenance or restoration.
2. Seeing that water is a valuable public property owned by the common people, the appropriate use of water must be ensured so that all of the people can enjoy its blessings in the future.
3. On the occasion of the use of water, attention must be paid to the avoidance or minimization of an influence on the sound watery rotation and its maintenance.
4. For any phenomenon occurring in the course of the water cycle that produces an effect in a subsequent cycle, the water must be controlled synthetically and completely in every river basin.
5. Seeing that the maintenance or restoration of the sound watery rotation is a task common to mankind, activities must be completed under international cooperation.

To understand the contemporary water problem, a holistic approach based on hydrology and the problem-solving dialogue of stakeholders should be required.

1.2.2 Public Awareness of the Water Crisis

The awareness of the water crisis has risen recently, and people have been searching for adaptations to various old and new water problems. People started to be aware of the water crisis when they experienced unsatisfactory water for living and social activities, both in quality and quantity. China has a famous proverb that "the one ruling the Yellow River rules the world." The meaning of these old words will be accepted in the present society as well. The "water festival" in Asian countries is a national event showing the prestige of the regime of the King. The proverb can be explained as "the one ruling the water rules a country," such as the "South-to-North Water Diversion Project" currently operated in China. This project carries the water in the southern part to the northern part in order to gradually reduce the water shortage, and it is a larger project than the Three Gorges Dam Project. In the short future, the proper management of water will be a strategic target in China due to its higher presence in the world.

Water crises like water shortages, floods, and pollution have a long history in many regions and people have been prepared in advance. The result of this is the prosperous urban and rich rural areas.

There is a question of how Japan currently perceives the water crisis now. In July 2009, the Japanese government produced a questionnaire entitled the "National Survey on Water Crisis Awareness" within the "National Traffic Administration Internet Monitor Questionnaire" (MLIT 2009). The definition of "water crisis" in the survey was "having difficulty accessing water because of water shortages, brackish water pollution, water quality accidents, and damage from earthquakes and weariness of drainage facilities such as pipelines, purification plants, intake facilities, and dams."

Among the results, the answers for "possibility of various water crisis occurrences" are shown below:

1. Over 80 % reported a water crisis because of earthquakes and weariness of facilities.
2. Over 70 % reported a water crisis because of water quality accidents, floods, and water shortages.
3. Over 50 % reported a water crisis because of brackish water.

The results of the questionnaire showed the close perceptive relationship between earthquakes and water crises, and the reason for this was possibly due to the Great Hanshin Earthquake in 1995, when people realized the importance of water. On the other hand, frequently broken drainage accidents also appeared to influence the awareness of the water crisis as well. Moreover, over 70 % of people realized that the water crises caused by water quality accidents, floods, and water shortages revealed a higher concern for living conditions.

Based on the results of the above survey, the answers for developing the necessary policies were as follows:

1. 75% of people suggested that policies should be in place to develop facilities that are resistant to earthquakes and weariness, followed by 66 % for the development of multi-route water supplies and storage facilities, and 43 % for the settlement of rain storage facilities. In this way, the survey provided for direct policy support for solving the water crisis.
2. The survey showed the awareness of preparation in advance by answering policies for individual water storage (43 %) and governmental water storage (40 %). However, the survey reflected less understanding for fewer publications among citizens on further detailed images and effects on water storage.

As the survey results showed, citizens realized that the water crisis existed and showed increasing recognition on hard and soft policy support.

1.2.3 Water Environmental Security

The solutions for water problems have always been an indispensable topic for survival from the emergence of humans, and it is the most urgent topic of water to be solved in the twenty-first century as earth environmental problems are getting worse. The key factors between water and global warming adaptations and lifestyles in human activities have been understood in a new vision. The UN General Assembly held in November 1980 formed the concept of the "International Drinking Water Supply and Sanitation Decade (1981–1990)," which was aimed to popularize drainage system technologies to improve the sanitation service in developing countries. Different from the problem of mass consumption of water in developed industrial countries, the water problems in developing countries are considered to originate from poverty, human rights problems, and survival problems. The 3rd World Water Forum held in Kyoto, Shiga, and Osaka in 2003 discussed the marketing and privatization of water and the security insurance of water. The most important consideration when considering the use pattern of water must be insuring the security and stability of water systems in living spaces, urban spaces, and industry. In other words, the social structure makes the system easily disrupted when the stability is disturbed. To prevent the disruption, sustainable water resources environment policy and international environmental cooperation are urgently needed to establish the security of the water resources environment. The first Asia Pacific Water Summit was held in Beppu City, Oita prefecture, Japan, on December 3–4, 2007. The theme of the summit was "Security Insurance of Water: Leadership and Responsibility."

"Water resources and environmental security" (or "water environmental security") is a concept brought to overcome the serious facts of the water crisis. Water for lives, water for living, water for urban industry, and water for earth environment are experiencing a crisis due to global warming, acid rain, and desertification that affect the recycling of water, and thus the water amount and water quality are reaching dangerous levels. The reservation of water resources and conservation of water environments were completed by technological measurements in the twentieth century. In the twenty-first century, the most important topic is to strategically work so-called water security on the water crisis from technological, social, international, and earth-based views. If a narrow sense of definition exists for water security, it shall be "fundamental and institutional preparation of water environmental security in urban spaces."

Water environmental security is an immature concept. The security insurance of humans started being discussed at the end of the twentieth century in international conferences and UN conferences. For example, "security insurance of urban area" and "security insurance of society" were the themes extended from the past discussions. The term "sustainable development" appeared in 1987 in the report "Our Common Future" by the World Commission on Environment and Development (WCED). Until now, this has been the basic concept of many environmental policies and has been much discussed and applied. The concept of sustainable development likely requires an understanding of the conceptual effectiveness discussed previously and the establishment of methodology before it can be accepted by the society of water environmental security.

Water environmental security is closely related with sustainable water resources environmental development and international environmental cooperation. Water resources environment problems cannot be merely discussed from the concept of the hydrologic water cycle or the "human and water culture" of regional sociology. To extend the conservation of the water environment to water resources and environmental security, a new policy framework should be built by working on new topics such as "commercialized water" and "marketing water" that have shown up in the twenty-first century.

Water environmental security is an outlook for sustainable water resources and environment policies, not only in Japan but also in the Asia Pacific region. The reason is that the Japanese economy is expected to grow by cooperating with Asian countries. Therefore, it is necessary to observe the method of international cooperation, which is the axis of the water resources environment.

1.2.4 Water Security for Disaster

The Tōhoku earthquake occurred on March 11, 2011, and attracted the fear and importance of water since the tsunami and nuclear plant accident. Transferring from the ideal "water" that consists of water for lives to nonideal "water" that includes water that brings disaster has caused people to change their perception. In this way, the importance is understood in a new vision of the key factors that water is involved in the growth of the human population, as well as human activities such as food and lifestyles. On the other hand, the survival of humans cannot be split by water. In the twentieth century, negative heritages left by development and wars started to become risks for survival. Now, all of those risks in earth, social, and human systems are pushed in front of us. In order to correctly understand the water crisis to develop the proper tools for problem-solving, incorporation of strategic visions in addition to the adaptations to global warming is important, instead of merely focusing on past problem-solving policies. The goal of the security insurance of water is to prevent the disruption of water systems in lifestyle and urban structures. This disruption is much larger than the existing climate change definition.

A "third way" has been searched for during the transaction of social water resources development enterprises, which cope with the expansion of traditional agricultural systems, industrial water consumption, and domestic water consumption. Since the influence of climate change on the water resources management has exceeded the assumed level, the establishment of the "security insurance of water" is required to apply strategic adaptations and realize sustainable development.

In the Human Development Report 2006 "Beyond scarcity: Power, poverty and the global water crisis," was identifed as severe water problems after the UNCED (UNESCO 2006). In the report, two topics were focused on. The first topic of "water for life" focused on providing secured water supply, drainage, and sanitary facilities. The second topic of "water for living" referred to the problem that governments are facing to manage water fairly and efficiently as a common resource, both domestically and internationally. This report was special since it indicated that the deficiency of the global water crisis was not just the physical amount of water but also the rooted privilege, poverty, and injustice surrounding the use of water. Thus, the obvious unfair usage of secured water and sanitary facilities was stressed upon. The report also indicated that cities with high incomes in Asia, Latin America, and Southern Sahara Africa have public water companies that supply hundreds of liters of water per day at a low price for their citizens. However, in the slums or rural regions in the same regions, only 20 l per day per capita could be kept to meet the basic needs of humans. Moreover, women and girls must sacrifice their time and education to obtain water, which is a double disadvantage in these regions. This report referred not only to the conflict between developed industrial countries and developing countries but also to the enlarging economical gap in developing countries. In addition, the world has enough water for domestic use, agriculture, and industrial use. The problem is that people living in poor regions have been removed from the public policy that restrained the infrastructure for the water supply, as well as the legislation privilege that helped them to sustain their lives. Therefore, the water problems may also become social problems or even political problems. To help solve these problems, an approach referred to as the security insurance of water has been developed as a new and restricted framework for water problems. The report defined the concept of "water security" as ensuring that every person has reliable access to enough safe water at an affordable price to lead a healthy, dignified, and productive life, while maintaining the ecological systems that provide water and also depend on water. When these conditions are not met or when access to water is disrupted, people face acute human security risks transmitted through poor health and the disruption of livelihoods.

When thinking about water problems, it is becoming important to discuss the "water security" itself by regional conditions.

1.3 Possibility of the Resurrection of IWRM

1.3.1 Background of IWRM

During the years 1950–1960, large dams had been constructed through water resources projects in order to recover and redevelop the social infrastructure after World War II. In 1970, environmental impacts from the construction of dams became strongly apparent and were subsequently criticized by the residents. This resulted in changes to the social movement to recover the broken ecosystem, and thus water resources development planning is converted from a demand-oriented approach to a comprehensive river basin management system.

Reflecting on this extravagant water development, an environmental preservation-oriented water management approach was considered. Moreover, the proposed new concept of "sustainable development" in 1987 led to the integration of management toward IWRM including natural ecosystems, economy development, and social participation.

Chapter 18 of Agenda 21 entitled "Protection of the Quality and Supply of Freshwater Resources: Application of Integrated Approaches to the Development, Management and Use of Water Resources" was adopted by the UN Conference on Environment and Development (UNCED) in 1992. The objectives of IWRM listed in 18.8 are based on the perception that water is a natural resource and integral part of the ecosystem that contributes to the improvement of society and the economy. The quantity and quality of water determines the nature of its utilization. To this end, water resources have to be protected, taking into account the functioning of aquatic ecosystems and the continuous cycling of the resource, in order to satisfy and reconcile the needs for human activities. In developing and using water resources, priority has to be given to the satisfaction of basic needs and the safeguarding of ecosystems.

Moreover, 18.9 further suggested that IWRM, including the integration of land- and water-related aspects, should be carried out at the level of the catchment basin or subbasin. Four principal objectives should be pursued as follows:

(a) To promote a dynamic, interactive, iterative, and multi-sectoral approach to water resources management, including the identification and protection of potential sources of freshwater supplies that integrate technological, socioeconomic, environmental, and human health considerations
(b) To plan for the sustainable and rational utilization, protection, conservation, and management of water resources based on community needs and priorities within the framework of national economic development policy
(c) To design, implement, and evaluate projects and programs that are both economically efficient and socially appropriate within clearly defined strategies, based on an approach of full public participation, including that of women, youth, indigenous people, and local communities in water management policy-making and decision-making
(d) To identify and strengthen or develop, as required, in particular in developing countries, the appropriate institutional, legal, and financial mechanisms to ensure that water policy and its implementation are a catalyst for sustainable social progress and economic growth

The Millennium Development Goal of halving the total number of people who are unable to reach or to afford safe drinking water by 2015 was developed at the 2nd World Water Forum (WWF2) and further stated the importance of IWRM and better water governance, given high-profile international recognition. The concept of water security was also introduced at WWF2, and this involves the sustainable use and protection of water systems, the protection against water-related hazards (floods and droughts), the sustainable development of water resources, and the safeguarding of access to water functions and services for humans and the environment.

1.3.2 The Definition of IWRM

The term IWRM has been defined by the Technical Committee of the Global Water Partnership (GWP 2000) as a process that promotes the coordinated development and management of water, land, and related resources to maximize the resultant economic and social welfare in an equitable manner, without compromising the sustainability of vital ecosystems. The UN has defined IRWM as a step-by-step process that takes time. By responding to changing social, economic, and environmental needs or impacts, one can gradually achieve better and sustainable water resources management as if moving up a spiral, through such means as progressively developing water resources in a basin, building a more integrated institutional framework, or improving environmental sustainability.

The US Army Corps of Engineers Civil Works Directorate summarized the principles of IWRM from a spectrum of international, federal, state, and nongovernmental organizations (NGOs) (USACE 2014). The principles shared the common notion to seek sustainability, balance economic, environmental, and/or social outcomes and promote economic development and the protection of ecosystems without harm to quality of life, safety, or security through:

1. Holism: adopting a watershed approach and perspective to look at the interconnections among local water issues and broader regional or watershed issues
2. Integration: integrating water resources planning and management with that of other resources (human, natural, and financial) for a balanced approach
3. Collaboration: collaborating with all stakeholders whether governmental, institutional, business, or the public
4. Participative decision-making and collaborative modeling: engaging stakeholders throughout a planning/decision-making process and tools that help visualize the impacts of management decisions on the watershed system
5. Sound science and innovation: leveraging the best available information, processes, and tools to support decision-making
6. Adaptive management: recognizing that in light of uncertainties, management is a dynamic process that requires ongoing monitoring and adaptation to changing conditions

1.3.3 The Limits of the IWRM

The history of IWRM is categorized into four different periods:

1. 1800–1945: development of a simple object to water demand for regional economic development; cost-benefit analysis is a powerful decision-making tool for evaluation of the projects.
2. 1945–1992: harmonize multiple objects in environment and development; environmental impact assessments were added for water development projects.

3. 1992–2015: seek a more effective way to develop IWRM; multidimensional factors and values were investigated.
4. 2016–2030: resolution of critical water incidents toward solutions; SDGs goal will be examined.

The concept of IWRM has been in continual development for many years, and thus many critiques have been generated over this time (Hamasaki et al. 2012). The critique of Prof. Dr. Asit K. Biswas (2004, 2008) pointed to the nonexistence of meetings regarding IWRM, which has likely contributed to the lack of success. Over the past 60 years, there has been no paper that has carefully assessed IWRM. Other researchers have published critical papers related to the controversial points of IWRM (Allan 2003; Hornidge et al. 2011; Mosse and Sivan 2003; Mukhtarov 2009; Van der Zaag 2005).

1.4 Development of IWRM Toward a Sustainable Society

1.4.1 Construction of a Sustainable Society

In order to change the developmental economic policy on building a comfortable, secured, sustainable society, it is necessary to face severe situations and to have the ability to implement "unregretful policies." Redefining the significance of sustainable society by clarifying the concept and active instructions and subsequently implementing policies are highly required.

The key to looking ahead to the concept of sustainability is understanding the current world environment and social economy. The rapid changes of the earth environment and the accelerating globalization of the economy are causing the destruction of the social economic system and escalation of civil issues. Human crises have existed from the past and will continue to occur in the future. The world now was built by various cooperative systems as the foundation. Those systems were based on the new consensus system that included all the international standard systems of the UN, ISO, and WTO. This is the baseline of the present economic situation. Today's UN system has disturbed the peaceful world order and the economic aids produced by the Official Development Assistance (ODA) program, which has enlarged the gap between developed countries and developing countries. Consequently, the number of refugees and unemployed workers has increased, and an ever stratifying society was reached. To solve these problems in structure, the opposite concept of sustainable society must be realized.

Agricultural society as a stable and adjustable status of a social system has supported human lives, and it has been considered to be a "sustainable society" for thousands of years. With a certain population density and arable area, the climate ensures a certain crop yield and society adjusted according to the yield. However, there are limitations of that social system and of its objectives. The balance between the power of protecting traditional systems and the power trying to revolve the social systems came from inside the system. In other words, both international and domestic communities understood the limitation of society and people had to accept the variety of values. This was supposed to create a sustainable society for change, rather than maintain the stable sustainable society.

1.4.2 Origin of the Sustainable Society

In the Saxony region of Germany around the 1700s, the mining exploration industry was threatening the living environment of the people. The same deforestation caused by development of the mining industry is destroying people's living environment today. Carlowitz (1713) studied the situation and distinguished the problems while considering the social economic condition as the background. As a result, a concept that stressed the importance of forest conservation by solving the deforestation problems was developed. The author also proposed an epochal idea that the purpose of forest sustainability was to use forests at a rate that the forest could reproduce itself. By following this rule, the recovery capability protecting the forest is maintained and will not cause damage to the forest. This idea is considered to be the origin of today's sustainability studies, and the sustainable society-related knowledge can be extracted from this idea. Though 300 years have passed since this publication was written, the problems of forest conservation have still not been solved. Furthermore, the construction of a sustainable society is important, due to the great risk of earth environment degradation. Even though researchers cannot ensure that the earth's environmental problems will be solved after 300 years, it is still necessary to have sustainability studies publicized continuously.

1.4.3 Objectives of a Sustainable Society

The complex social problems come from considering a system where the nature of problems is related to many elements. Splitting the complex and extracting the detached elements to form the assumption are one way to observe the complexity. Sometimes, there had been few rules concluded from the observations, but with the processes, the ideas of the best or the better processes were brought forward.

Topics for now and for the future are the existing thinking patterns that emphasize the causality and stereotype causality of modern life. In order to construct a sustainable society, the social and technological imaginations toward the future are necessary from a consensus on every decision-making level in order to influence the future and the implementation of irreversible strategic decisions.

Sustainable development may be defined as the use of environmental resources under the conservation of ecosystems and natural conditions that are impartial both between generations and between the South and the North in order to overcome poverty. However, other than this controversial definition, an absolute definition of sustainable society does not currently exist. Only the objectives of a sustainable society are common, and these include the admission of finite environmental usage, the restriction of sudden actions, and the realization of social justice.

The attitude of creating a sustainable society includes not predicting the future but searching for the potential of creating the future. The development of highly informed societies has revealed this sense of value. Wide usage of the internet transfers material things and services into information and eases the access. Regarding the creation of future societies, there have been changes from the idea prediction thinking to strategy-based thinking toward future-like scenarios.

The principle of a highly informed society is to have a clear view on "possibilities." In other words, the principle is not to predict the "possibility" of development with experiences but to include technological revolutions that could ultimately increase the "possibility" and thinking paradigm transformations as the elements of "possibilities." Sustainability in the highly informed society refers to the elimination of information "poverty" by changing the current society system with communication and by assuring the justice of accumulative information between generations.

Sustainable societies, including the sustainability in our experienced agricultural society, industrial society, commercial society, and informative society, should have a new start and objective. The objective of a sustainable society is to maintain the social economic system's concept and action that aim at solving coming problems.

1.4.4 Local and Global Sustainability

Local sustainability refers to problems that could be solved by local resources and efforts. This concept started with differing problems that could not be solved by local powers, which was the transforming period from pollution problems to environmental problems. When the pollution problems were being discovered as a social problem with a clear victimizer and victim relationship, the ambiguity of environmental problems was brought forward. However, in addition to the detailed pollution and causal problems, local environmental problems that affected our living had also been known. Therefore, it was realized that if the local pollution was profoundly perceived and solved, then the local problems could be eliminated. However, the involvement of national and local administration is often required to solve the pollution problems between the confronted enterprises and the local people.

The local sustainability is not ensured by merely solving the pollution problem, as it is also important to improve the amenity while solving pollution problems. The proper focus of the local sustainability is not only to understand the pollution in degraded regions but also to perceive the fact of disappearing native social economic and cultural history.

Global sustainability has targets at the earth level. Without solving all of the problems, human and other species are at risk for extinction. To achieve these targets, international cooperation and treating the earth as a public asset are needed. Obtaining the sense of policy and vision on reconstructing the earth system and repairing the relationship requires a "creative attitude," which refers to the search for the potential of a future, rather than predicting the future. Based on this "creative attitude," the possibility of developing action patterns in a highly informed society can be explored (Nakagami et al. 2015).

1.4.5 Water Sustainability

Sustainability became a hot topic after the publication of Our Common Future by Brundtland (1987) and has gained applications in fields such as policy-making and management strategies. In the 1990s, its application expanded to much broader fields at different levels, such as economic sustainability, environmental sustainability, and even more specifically to water resources sustainability. The sustainability of water resources refers to the water resources system that is designed and managed to fully contribute to the objectives of society, now and in the future, while maintaining their ecological, environmental, and hydrological integrity (Loucks and Gladwell 1999). This concept indicates that sustainability relates to more than one aspect and is dynamic, rather than static. Therefore, the methods for realizing water sustainability should take these aspects into consideration. Savenije and Van der Zaag (2008) developed a conclusion regarding the development of IWRM and stated that it contains three of four dimensions, including natural side, human side, and different spatial and temporal scale variations. Under this condition, Bedrich et al. (2012) discussed that a wide range of indicators helps to quantify the sustainability and ease the measuring process to the ideal sustainability target. The indicator-based assessment method is widely used for the evaluation of sustainability including water sustainability. The Water Poverty Index (Sullivan 2002) and Watershed Sustainability Index (Chaves and Alipaz 2007) are examples of global water sustainability indexes. As for regional indexes, the Canadian Water Sustainability Index (Policy Research Initiative 2007), West Java Water Sustainability Index (Juwana et al. 2010), and the Watershed Sustainability Index (Ana et al. 2012) represent the fitness of this method for measuring sustainability.

The indicator-based method indicates the "integrity" characteristic of the sustainability concept. Nevertheless, the dynamic movements over the long-term are not reflected at all. Fortunately, studies concentrating on long-term performance have come up with a new tool, which is the scenario-based method. Since water shortage problems are occurring around the world, scenarios of water supply and demand have been developed. A global-level study designed two scenarios from the year 1990 to 2025, based on different irrigation efficiencies, and the results showed that the improvement of irrigation effectiveness could save approximately one-half of the additional water resources (David et al. 1998). Another paper was conducted at a national-level in Egypt (Mohamed 2007), with the conclusion regarding the challenges and opportunities for Egypt from the year 2000 to 2020. This research designed three scenarios based on different development speeds, and the results showed that water policies that were developed at a faster speed promoted the effectiveness of water systems and could lead to a sustainable future.

1.5 Evaluation of Water Resources Development Projects

1.5.1 Objective of the Water Resources Development Projects

The objective of water resources development enterprises is to manage the reconstruction of water resources development or usage facilities and the facilities operations and to ensure a stable supply in industrially developed regions and crowded areas. The evaluation of current Japanese enterprises is operated based on this objective. The evaluation considers whether the enterprise has a high priority in meeting public needs and the public profits. The Ministry of Land, Infrastructure, Transport, and Tourism evaluates enterprises based on an adoptive evaluation of new enterprises (when new enterprises are adopted, a cost-effectiveness analysis is operated as an evaluation step), reevaluation (during the period after the enterprises are adopted and before the construction, reevaluation is performed to make the necessary corrections or to suspend some enterprises), and post-evaluation (after the enterprises are completed, an evaluation is performed to confirm the effectiveness and environmental influences, in order to prepare the improvements or discuss the plan and research methods for the same categorical enterprises). The contents of the comprehensive evaluation include cost-effectiveness analysis that are sometimes highly affected by political conditions. The time required to go from the concept phase of a water resources enterprise to the planning, construction, application, maintenance, and disposal phases may be lengthy, and so it is difficult to evaluate the water resources development as a single unit over the short term.

Entering the twenty-first century, the current evaluation of river security is changing since the annual maximum precipitation changes in relation to climate change. On the other hand, a water resources development enterprise led by the government was constructed on May 28, 1997. The enterprise added equipment and conservation of rivers and introduced a river improvement plan according to regional wills after the proposed Act on revising of River Law. The amendment was the first revision of the old River Law since 1896, and it changed the future of water resources development enterprises. By adding flood prevention and water usage into the river environment as the objectives of river management, the idea of reflecting the regional wills as the consensus for social river management shall be kept in place for the next 100 years. In this way, a new evaluation method for water resources development enterprises is required. The over 100-year duration to realize the water resources development enterprises from the imagination to completion requires a sustainability assessment based on the development of the regional society and rapid changes of the earth's environment.

1.5.2 Sustainability Evaluation of the Water Resources Development Project

A sustainability assessment is based on the goals of water resources development enterprises by selecting economic, social, and environmental elements as evaluation indicators. The suitable sustainability assessment needs clear objectives of the operating evaluation. In the past, water resources development enterprises were performed to increase social benefits, to change the natural environment, and to reform the social system. These are the normal elements when considering the economy, society, and the environment. However, these are difficult to realize in the evaluation of multi-aspect projects, and so a cost-effectiveness analysis methodology was proposed in 1844 as a simple and persuasive technique.

A total of 170 years have passed since the cost-effectiveness method was proposed, and it has been challenged to obtain trustworthy information during evaluation. In 1973, the US Water Resources Council (WRC) set four essential elements for effectiveness analysis based on the expansion of public enterprises and social evaluations. They included the development of the national economy, development of the regional economy, improvement of the environment, and improvement of social benefits. The application of cost-effectiveness on water resources development enterprises was not widely accepted, but it did provide a path for sustainability assessment with the comprehensive involvement of the economy, society, and the environment.

Mr. Daniel Beard, the president of US Bureau of Reclamation (USBR), presented a report entitled "The End of the Dam" to the International Commission on Irrigation and Drainage (ICID). The report did not appear to affect the whole world, though it spreaded the policies of the main enterprises in the USBR such as transforming from the construction of dams and waterways to the recovery of the environment or saving energy (Beard 2003). From the nineteenth century to the end of the twentieth century, dams have stopped being used as the instrument of irrigation that is the foundation of agricultural development. However, the effect of water resources development enterprises like dams to the social economic condition and the development of public enterprises policies from the origin of the problem statement are important to understand. As well as Japan, the problem statement started from the transformation of water resources development policies and river administration.

The goal of the USBR has changed from dam construction to sustainable water resources management and has contributed globally to the understanding of energy savings and the recovery of the environment. At the same time, it has influenced the Japanese water resources development enterprises.

The basic concepts of water resources development enterprises, including the improvement of living levels and of water security that meets the social economic development needs, were reconsidered and a pragmatic solution was needed. The purpose of the statement by Daniel Beard was to create a new way to confront the powerful large-scale water resources development enterprises. The high development and diversity of social economic systems result in water resources environmental problems. Based on this, the past improvement of the water supply security plan was not enough, and the limited public understanding of the water resources development plan that was reached made it difficult to have a social consensus. Therefore, the enlargement of the water resources origin problems and urban problems by ignoring these questions is unacceptable when implementing the comprehensive water resources resolution.

As the evaluation elements of water resources development enterprises, economic, social, and environmental efforts are needed, while "social consensus" and "environmental harmony" are essential as well. The evaluation of sustainable water resources management involves the transformation from the past economic evaluation of agricultural effectiveness to the sustainability assessment.

The World Commission on Dams (WCD) published a reported entitled "Dams and Development: A New Framework for Decision-Making" in November 2000 (World Commission on Dams 2000). The Commission grouped the core values that included its understanding of the effectiveness of future decisions under five principal headings: equity, efficiency, participatory decision-making, sustainability, and accountability. Moreover, seven strategic priorities that were supported by a set of policy principles provided a principled and practical way forward for decision-making: (1) gaining public acceptance, (2) comprehensive options assessment of dams and their alternatives, (3) addressing existing dams, (4) sustaining rivers and livelihoods, (5) recognizing entitlements and sharing benefits, (6) ensuring compliance, and (7) sharing rivers for peace, development, and security.

The comments on this report were diverse from different positions, and it can be regarded as the guide for water resources development in the twenty-first century. These core values and strategic priorities can easily pass for not only public works but also national land conservation.

During the First World Water Forum held in Mar del Plata in March 1977, the Mar del Plata Action Plan was enacted including (A) assessment of water resources; (B) water use and efficiency; (C) environment, health, and pollution control; (D) policy, planning, and management; (E) natural hazards; (F) public information, education, training, and research; (G) regional cooperation; and (H) international cooperation. The concept based on the public acceptance of water resources development enterprises was concluded from the review and improvement of this plan, which is considered to be the axis of water management. The framework of the Mar del Plata Action Plan is the foundation of water resources management plans and is the base of water policy-making. Based on this action plan, the core values from the WDC (Water Development Council) will be questioned in future water resources development enterprises as well. Therefore, with the doubts concerning the existing cost-benefit analysis and cost-effectiveness analysis, an alternative evaluation to "the third way" is urgently required.

Although holistic approaches may be used as the framework of sustainability assessments to solve the water problems, the methodology and standards are not clear yet. For example, the vision of sustainability assessments on water resources development enterprises, how to involve the complexity of the problem, the large scale of enterprises, and incorporating public acceptance into assessment are topics for further studies.

1.6 New Perspectives of IWRM

1.6.1 New Trends of IWRM

The OECD proposed principles for water governance during a session at the WWF7 in Korea, and the purpose of these principles were to promote efficiency and effectiveness in water management.

The session "Principles on Water Governance: From Vision to Action" was chaired by the OECD's Secretary General and provided a platform for multistakeholder experience sharing on the building blocks of good governance covered in the 12 principles: (1) multilevel governance; (2) managing water at the right scale; (3) policy coherence; (4) capacity; (5) data and information; (6) governance-financing nexus; (7) regulatory frameworks; (8) governance and innovation; (9) integrity and transparency; (10) stakeholder engagement; (11) equity across users, people, and places; and (12) monitoring and evaluation.

Moreover, the World Water Council (WWC) reported "Integrated Water Resource Management: A New Way Forward" at the WWF and added additional factors for new directions of IWRM. These reports oriented the effective strategies for dynamically catalyzing and managing changes at all levels, such as the facilitation of processes for social learning that are supported by data, communications, and empowerment to solve problems and learn by doing, which work with and reinforce reform processes and investments. They also included operating mechanisms that bridge the strategy setting and problem-solving platforms to enable sectors and stakeholders to come together to negotiate, coordinate, collaborate, and jointly innovate.

1.6.2 Sustainable Development Goals and Water

The International Decade for Action "Water for Life" 2005–2015 was developed to resolve water issues and sanitation. Based upon the experience and know-how obtained from these activities, the open working group at the UN submitted a goal to ensure access to water and sanitation for all, which was finally adopted by the UN Assembly on September 25, 2015.

The content of Goal 7 is as follows:

1. By 2030, achieve universal and equitable access to safe and affordable drinking water for all.
2. By 2030, achieve access to adequate and equitable sanitation and hygiene for all and end open defecation, paying special attention to the needs of women and girls and those in vulnerable situations.
3. By 2030, improve water quality by reducing pollution, eliminating dumping and minimizing the release of hazardous chemicals and materials, halving the proportion of untreated wastewater, and substantially increasing recycling and safe reuse globally.

4. By 2030, substantially increase the water-use efficiency across all sectors and ensure sustainable withdrawals and supply of freshwater to address water scarcity and substantially reduce the number of people suffering from water scarcity.
5. By 2030, implement integrated water resources management at all levels, including through transboundary cooperation as appropriate.
6. By 2020, protect and restore water-related ecosystems, including mountains, forests, wetlands, rivers, aquifers, and lakes.
7. By 2030, expand international cooperation and capacity-building support to developing countries in water- and sanitation-related activities and programs, including water harvesting, desalination, water efficiency, wastewater treatment, recycling, and reuse technologies.

1.6.3 New Paradigm of IWRM

The largest impact factor related to IWRM is climate change. The influence of climate change on water issues became clear by the International Panel on Climate Change Fourth Assessment Report (IPCC AR4, WGII, Chapter 3). From the standpoint of IWRM, the importance of the management of water crises has rapidly increased. The adaptation plan for water resources is limited by traditional measures, and so the strategic adaptation methods for water management should be examined.

The record of observations related to climate change and its influences are as follows (IPCC 2007):

1. A rising trend 0.74 °C of air temperature during the past 100 years (1906–2005).
2. The global mean sea water level increased by 1.8 mm/year from 1961 and 3.1 mm/year from 1993.
3. Precipitation in the eastern area of North and South America, Northern Europe, and the northern and middle areas of Asia has generally increased every year from 1900 to 2005.

To cope with the present water crisis due to climate change, the subjects of IWRM are summarized as follows (Nakagami 2009):

1. The purpose of IWRM is to manage surface water, groundwater, and virtual water as part of the integrated water resources.
2. To execute the IWRM, sustainable and comprehensive concepts and approaches should be considered to resolve various problems.
3. Various interests will occur once the IWRM is implemented. More possible interest groups emerge and should participate in the decision-making process.

4. The IWRM should be managed to take strong a stand on the water users and be equitable.
5. To examine the IWRM, interdisciplinary approaches should be implemented that aim to integrate knowledge and information from various scientific fields.

1.7 Conclusion: Possibility of Policy Science as Design Science

"Problem discovery" exists from the view of policy science. How do we usually discover problems, and how can we discover the nature of the problems?

It has taken a long time to profoundly understand the potential issues associated with pollution problems. Shōzō Tanaka in the Ashio Copper Mine issue and Minakata Kumagusu in the combination of shrines issue – nature conservation campaign – are worth being respected for their prediction of "problem discovery" more than 100 years ago. The strong thought before the policy became a science should be the real "problem discovery." Chlorofluorohydrocarbon was considered to be the greatest invention in the twentieth century but has since been found to be dangerous, and the irreversibility of global warming has been indicated. Thus, human concerns and other scientific considerations are important.

The discovery of problems cannot ensure the solution of problems, but as long as the problems are not found, there will be no solutions. Decision-making in the current society is complicated as it involves diverse stakeholders. In this case, a number of discussions must determine conflicts of interest during the decision-making processes.

Confirming the development direction of policy science requires considering the earth, society, and human system as a complex social problem since now there is no detached existence. It is also necessary to solve the problems based on the strong will of strategic decision-making, as well as the cooperation based on tolerant understandings. The role of strong will is policy science and is something that is essential now. The challenge is not simply discovering and solving the problems, but analyzing current topics and forming creative thinking for the future.

Acknowledgments This work was supported by the Research Institute for Humanity and Nature (RIHN), "Designing Local Frameworks for Integrated Water Resources Management" Project (C-09-Init), and Ritsumeikan University, Ritsumeikan Global Innovation Research Organization (R-GIRO).

References

Allan JA (2003) Virtual water – the water, food, and trade nexus useful concept or misleading metaphor? IWRA Water Int 28(1):4–11

Ana EC, Ricardo O, Nicole K, Henrique C, Guido S, Manuel S, Jaime A, Jorge O, Tobias R, Michelle S, Hugo M (2012) Application of the watershed sustainability index to the Elqui river basin, North-Central Chile. Obras y Proyectos 12:57–69

Bedrich M, Svatava J, Tomas H (2012) How to understand and measure environmental sustainability: indicators and targets. Ecol Indic 17(4–3):5–13
Biswas AK (2004) Integrated water resources management: a reassessment. Water Int 29(2):248–256
Biswas AK (2008) Integrated water resources management: a reassessment. Water Resour Dev 24(1):5–22
Chaves HML, Alipaz S (2007) An integrated indicator based on basin hydrology, environment, life, and policy: the watershed sustainability index. Water Resour Manag 21(5):883–895
Daniel PB (2003) Changing Japan's river management policy. Retrieved from http://www.mm289.com/RPN/1192/biard.html
David S, Upali A, David M, Radhika d S, Randolphr B (1998) World water demand and supply, 1990 to 2025: scenarios and issues, Research report 19. International Water Management Institute, Colombo
Hamasaki H, Kato H, Nakagami K (2012) New perspectives of IWRM. In: Proceedings of the 12th conference of Society of Environmental Conservation Engineering, 3 Sept 2012
Hornidge AK et al (2011) Reconceptualizing water management in Khorezm. Natural Resources Forum, Uzbekistan
IPCC (2007) Climate change 2007: the first working report. http://www.ipcc.ch/SPM2feb07.pdf
Juwana I, Perera B, Muttil N (2010) A water sustainability index for West Java-Part 2: refining the conceptual framework using Delphi technique. Water Sci Technol 62(7):1641–1652
Loucks DP, Gladwell JS (1999) Sustainability criteria for water resource systems, International hydrology series. Cambridge University Press, Cambridge
Masafumi ITO (2012) Urban river management in Japan. The 9th international forum on waterfront and watershed restoration, 24 November 2012
Ministry of Land, Infrastructure, Transport and Tourism (2009) Awareness survey of domestic water crisis
Mohamed NA (2007) Water resources in Egypt: future challenges and opportunities. Water Int 32(2):205–218
Mosse D, Sivan M (2003) The rule of water: statecraft, ecology and collective action in South India. Oxford University Press, New Delhi/Oxford
Mukhtarov (2009) The global policy network behind integrated water resources management: is it an effective norm diffusor? Ecol Soc 19(4):11
Nakagami K (2009) Global warming and IPCC. Fundamental and approach of college of policy science, 2nd edn. Minervashobo Com., April, 2009 (In Japanese)
Nakagami K'i, Wang S, Xiaochen C (2015) Sustainable water resources management of the Nile River, Egypt. Policy Sci 22(2):11–20
Savenije HHG, Van der Zaag P (2008) Integrated water resources management: concepts and issues. Phys Chem Earth Parts A/B/C 33:290–297
Sullivan C (2002) Calculating a water poverty index. World Dev 30(7):1195–1210
U. S. Army Corps of Engineers Civil Works Directorate (2014) Building strong collaborative relationships for a sustainable water resources future understanding Integrated Water Resources Management (IWRM)
Van der Zaag P (2005) Integrated water resources management: relevant concept of irrelevant buzzword? A capacity building and research agenda for southern Africa. Phys Chem Earth 30:867–871
von Carlowitz HC (1713) "SYLVICULTURA OECONOMICA" hauswithliche Nachrichit und Nature masige Anwesung zur wilden Baum-Zucht
World Commission on Dams (2000) Dams and development a new framework the report of the world commission on dams for decision-making. Earthscan Publications Ltd, London/Sterling
World Water Assessment Programme (2006) The 2nd world water development report: water; a shared responsibility. UNESCO, Paris. Available from http://www.unesco.org/water/wwap/wwdr/wwdr2/table_contents.shtml

Part II
Design: Designing Water Resources Management with Collaborative Activities of Multistakeholders

Chapter 2
Participatory Approaches to Environmental Management: Future Design for Water Resources Management

Hisaaki Kato, Ken'ichi Nakagami, and Malcom Cooper

Abstract Participatory approaches that feature the broad inclusion of local stakeholders have become a basic design requirement for future water resources management. However, the actual implementation of an effective management plan currently remains a challenge. In this chapter the authors focus on the concept of "system integration" in integrated water resources management (IWRM) as the root cause of the problem. More than ten years have passed since the publication of Asit K. Biswas' ultimately critical treatise, wherein he argued that although he could understand the necessity for IWRM, it cannot be reflected in management plans. Nonetheless, this problem continues to be neglected. We argue in this chapter that the problem exists in ways of thinking that are predicated on hard-path water resources management. In the conclusion, we discuss the outlook for the alternative soft-path, adaptive management, using an aborted dam project in Japan as a regional case study.

Keywords Participatory approach • Integrated water resources management (IWRM) • Adaptive management • Hard-path • Soft-path • Dam-based management

H. Kato (✉)
Research Institute for Humanity and Nature,
457-4 Motoyama, Kamigamo, Kita-ku, Kyoto 603-8047, Japan
e-mail: hisaaki@chikyu.ac.jp

K. Nakagami
College of Policy Science, Ritsumeikan University, 2-150 Iwakura-cho,
Ibaraki, Osaka 567-8570, Japan

M. Cooper
Graduate School of Asia Pacific Studies, Ritsumeikan Asia Pacific University,
1-1 Jumonjibaru, Beppu, Oita 874-8577, Japan

K. Nakagami et al. (eds.), *Sustainable Water Management*,
DOI 10.1007/978-981-10-1204-4_2

2.1 Introduction

Water resources management typically takes a "hard-path" as opposed to a "soft-path" approach (Amory 1977). This approach covers various scales, ranging from integrated watershed development to the community level. In configurations of the local scale, community-level participatory water resources management is generally portrayed as the ideal management style, and to this day, global development agencies continue to offer support for its realization. The goals are to reduce government spending and enable the efficient utilization and independent management of water resources, through community participation in water resources management. However, few successful examples exist (Mukherjia et al. 2012). Furthermore, as Svendsen et al. (2000) point out, by raising the issue of "Second-Generation Problems," upon the installation of participatory management, a new series of problems can arise. For example, irrigation associations become dysfunctional, and/or collections for irrigation expenses completely fail within the regional community system, for those receiving aid implement plans in conditions that lack the capacity for development.

These problems not only occur in the limited area of irrigation management but are also common to water and sewer service maintenance and many water resources management projects. Further, an overview of international discussions suggests that the basic problematic point is that, unlike hard-path approaches that center on infrastructure, soft-path approaches that cooperate with a broad range of stakeholders stumble in the social implementation phase (Maksimović and Tejada-Guibert 2001). In this way, the conceptual transformation leading to a practical implementation of the soft-path approach stopped halfway and has remained unchanged since 1992, when an outline for national transformation from development toward environmental conservation was introduced at the United Nations Conference on Environment and Development. Still, academia and development organizations alike continue to be embroiled in debates around fundamental questions, such as "the meaning of water management" and "the problem of actual policy implementation."

From a policy perspective, this approach tries to integrate every kind of issue in complex situations, while regulating the structural elements of the budget, principle of conduct, organizational structure, knowledge gaps between stakeholders, the entirety of infrastructure including task environment, and so on. The consideration of land resources in addition to water resources has further problematized organizational structures, the structure of transmission of intention, the commitment of those participating, objectives shared by the regions that contain the organizations, perception gaps between stakeholders, infrastructure (task environment), and actionable capital (socioeconomic structure) in various regional units. The general idea of integrated water resources management (GWP 2000) attempts to synthesize these concerns, scientifically describe diverse realities, and take into account various water management knowledge and methods by generalizing them. However, a history of attempting to integrate all these diverse issues invites criticisms, such as "the importance of the concept can be understood, but it cannot influence actual management plans" (Biswas 2004, 2008). In other words, the problem is that although the theoretical aspect of the design is excellent, the actual implementation of the management plan is rather difficult.

IWRM has thus fulfilled an important role as a sought-after ideal, but it does not show any sign of being able to bridge the chasm between international community support for the idea and its actual implementation by society. As a result, reevaluating IWRM from the local level up has been an international challenge for the last several years. The current authors have a history of addressing this issue in Indonesia (Hamasaki et al. 2012). However, from experience gained focusing on developing countries, the authors are fully aware of the "limits of any approach that integrates several themes—each of which was traditionally regarded as a supreme theory of water management," as well as the need to examine the problems faced by developed countries, which have been modeled on such an approach. Until today, the main point of IWRM has been to allow consideration of the ways in which developed countries might share their successes and help to establish excellent water resources management systems in developing countries. Yet, considering the above problems faced by developed countries, it is necessary to attempt to create a new notion of values in IWRM.

2.2 Integration and the Supporting Stationarity Principle

When we look at the history of water resources management, we can see that IWRM has focused on measures crafted to accommodate water demand brought on by socioeconomic growth, flood control policies based on dams, and policies for remediation of water environments destroyed by industrialization and urbanization. However, as described in the introduction, implementation of IWRM has raised many problems. In recent years, its implementation has received a great deal of attention in international meetings, though the problem of how to take concepts to the level of planning and transcend the difficulties of the level of actual policy has been debated for years. Fully exercising his authorial position as an expert in water resources management policy, Asit K. Biswas (2004) identified the dysfunctional aspects of IWRM in this condition, adopting a reformer's style to discuss the form IWRM ought to take, what should be integrated, and so on. On the one hand, the impact of pointing out such problems was enormous; however, it is difficult to agree with the intentions and determine whether the contents of the critique were correctly perceived. Rather, having failed to resolve the challenges and proceed, a tautological situation is emerging.

Nevertheless, in real life IWRM continues to be evoked in many countries' public policies, as well as in debates about international water resources management. In the creation phase of water resources management plans, the simultaneous incorporation of IWRM policy formulation is an indispensable item for development aid organizations, like the Asian Development Bank, the World Bank, JICA, and so on.

This indicates a shift from highly abstract discussions over water resources toward debates over specialized efforts on behalf of the natural environment.

In this way, IWRM is a directional change, in response to the conditions of the age, and has the idiosyncratic quality of continuously surviving as a system. However, another characteristic of IWRM is the continued incorporation of a hard-path foundation that is primarily interested in infrastructure maintenance. This is made clear by references to the 1933 Tennessee Valley Authority (TVA), which was an example of IWRM from the outset (Snellen and Scherevel 2004). The TVA was not only a singular exemplar for the United States but also a global role model for hard-path style water resources development.

In this context, modern water resources management is the principle that takes all relevant elements of the entire river basin and the development of water resources management facilities as criteria and "integrates" them (Nakagami 2015). Taking an example from Japan, the Lake Biwa Integrated Development Project, which existed from 1972 to 1997 at a cost of approximately two trillion yen (US$16 billion), incorporated not only flood control and irrigation but also regional infrastructure maintenance and represented the integration of all forms of hard-path elements. However, because of excessive social costs and an overemphasis on hardware, this integrated river basin development project, in keeping with the times, created many environmental problems.

However, when dysfunction and the original need for IWRM were being identified, it was rare for questions like "Why is excessive integration starting from hardware even possible?" to be asked. Moreover, the mechanisms underlying such principles were scarcely points of inquiry. As Milly et al. (2008) point out, the reasons for this can be found in the assumption of "stationarity" currently underlying water resources management in developed nations. Stationarity has been positioned as an important basis for the promotion of a plan, once infrastructural facilities, such as dams, have been constructed.

Starting from a limited understanding of the natural environment as being something that fluctuates within unchanging outer layers, stationarity is the principle that maintains that perturbations in the river basin can be modulated by means of infrastructure. Using this principle, the design and production of general infrastructure, such as dams, have been based on the idea that the probability density function (PDF) of the yearly cycle between the annual flow rate and flood peak has characteristics that can be measured from observation instruments. However, because of climate change and so on, it has become difficult in recent years to manage even the hydrological variables in accordance with the way of thinking based on this kind of stationarity and the originating hardware situation.

Thus, Reuss (2003), Milly et al. (2008), and others used an interdisciplinary approach, based on simulation optimization, to reevaluate the efforts of the Harvard Water Program (Maass et al. 1962), for example, touching on efforts to produce estimates of non-stationarity and uncertainty through modeling. Against the background thus identified, much like the problems with public architecture in the United States raised by Ehrlich and Landy (2005), a need exists to renovate crumbling public utilities, including water-related infrastructure, in most of the developed countries that have traditionally been leaders in water resources management. It is for this reason that new strategies, which pick up where traditional stationarity left off, are necessary. However, this may merely be a repetition presented in a more modern style.

2.3 Deadlock on the Integration Principle: The Case of the Discontinued Niu Dam Plan in Japan

When thinking about regional water resources management, it is clear that attempting to move away from the issue of "integration" will not be easy. However, whether in developing countries seeking economic maturity or developed countries whose mature development means socioeconomic conditions of zero growth, in today's circumstances, the notion of integration itself is beginning to be seen as unreasonable. From a consideration of issues such as the "co-creation of science and society" (Lang et al. 2012), the authors of this article claim that "adaptation" is important in considering the society's future design, and an earnest confrontation with climate change and the myriad transformations of the natural environment is also a must.

Based on these claims, we take Yogo Lake, in Nagahama City, Shiga Prefecture, Japan, as a case study of existing IWRM, based on the diluted "design for the region's future" perspective, and discuss how the conversion from integration to adaptation can be realized. The subject of our study, Yogo-cho, Ika-gun in Shiga Prefecture, Shiga Prefecture's northernmost town, was incorporated into Nagahama City on January 1, 2010. The area of the region is 167.62 km^2, with 1245 households, and a total population of 3510; it is also the only region in the west of Kinki designated as a special heavy snowfall area. Seventy percent of the area is forest, and, until 1964, the main industry was charcoal production. However, after the main fuel sources shifted from wood, charcoal, and coal to petroleum, many people changed occupations, seeking employment in the factories of neighboring towns.

A look at the water environment reveals a dual system, with Lake Yogo connected to Lake Biwa, but also the uppermost stream of the Lake Biwa–Yodo River water system (the Takatoki River–Anegawa River Basin) (Fig. 2.1). When it was decided that Lake Yogo, once a closed system, was to be connected to Lake Biwa, in order to provide agricultural water, as part of the Kohoku Integrated Development Project (1966), it was clear that its ecosystems would be negatively impacted. Because of this, an artificial circulation system was created and improvements to water quality sought. On the other hand, a look at the layout of the river basin and Old Yogo Town shows how the river management plan was all but canceled because of the planned Niu Dam and "overcoming the hard-path" remains a problem.

Originally called the Takatoki Dam, according to the Department of Kinki Regional Development Bureau's (currently the Kinki Regional Development Bureau) 1980 plan, Niu dam would have been Japan's largest rock-fill dam at 145 m high, 474 m long at the crest, with a volume of 13.9 million m^3, and a reservoir

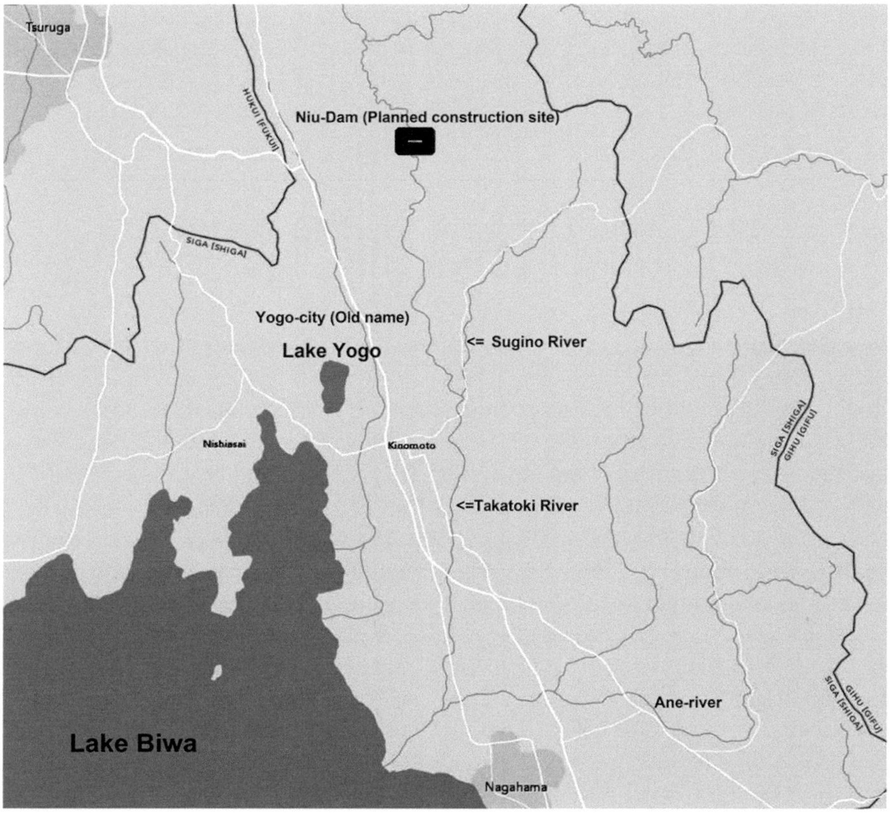

Fig. 2.1 Lake Yogo and Takatoki River–Anegawa River Basin

capacity of 150 million tons of water (planned value). In addition, this was the first time in Japan that a planned dam was canceled after the affected residents were relocated. The scope of the construction project and the fact that it was canceled after residents had been relocated are notable examples of the rethinking in Japan about the future of IWRM.

Further, at least in Japan, the multipurpose dams that have, until today, been the main solution to irrigation and flood control problems can no longer stand on the basic premise that there will be increases in the demand for water from agriculture in the future, and they are, therefore, no longer necessary for irrigation. In addition, with the rise of alternative civil engineering solutions, flood control can be sufficiently dealt with through river improvement projects. Nonetheless, problem-solving ideas centered on dams, based on traditional rational planning models, continue to occupy a considerable place in modern society.

However, because these projects begin on the presumption of "dam construction," they lack an appropriate policy in cases of project termination. Expressed another way, because of the large social cost of construction, configurations including other options, like discontinuation and alternative policies, are not considered.

Properly speaking, as a political policy, this obligates the creation of a tolerable level of alternative policies for dams and rivers, as well as several alternative options for problem-solving, as early as the planning stage. However, in the Niu Dam case, these types of risk aversion policies were not considered. Thus, the Niu Dam situation is not only a case of water resources management deadlock brought on by an initial insistence on modern mega-infrastructure in the hard-path style but is also one of the most important cases supporting consideration of the transition to a soft-path approach, through an appropriate combination of medium-to-small-sized facilities near the consumers of water resources rather than a mega-dam further away.

The main purposes of the Niu Dam were to provide for the demand for water resulting from the increase in population of the Kansai region and to provide thorough flood control for the Takatoki–Anegawa River Basin. Furthermore, as an integrated problem-solving method based on a dam, it epitomized management based on water storage. The project was originally planned to be concluded in 2010, and from the start of planning until 1996, the target goal of the relocation of 40 households was achieved. However, because of reduced water demand resulting from declining population in the Keihanshin region, the Kinki Regional Development Bureau and the Japan Water Agency finally decided to stop the plan, and it was canceled on January 16, 2014. The historical background to the termination of the plan shows that the proposals released in January 2015, by the Yodo River Basin Committee (established 2002) as part of their "Reconsideration of the Dam Construction Plan," had a large impact. While, in recent debates over IWRM, many skeptical eyes have been focused on management techniques that depend on storing water in large-scale reservoirs, and there are many who claim that "No Dam is a Good Thing" in line with a general trend toward dam removal, the discontinuation of the Niu Dam is not an event we can simply attribute to this trend.

The development of new water resources using reservoirs, as well as the uncertain economics of population decline and existing infrastructure, is important when considering the context of the burden they place on future generations. Dams are not resources that can be used indefinitely but rather are structures beset by problems from the beginning, such as the accumulation of sediment in the reservoir, the maintenance of the structure, and so on.

However, regions can be irreversibly altered as part of the preparation of a draft plan. In response to the proposed transformation of the region through the Niu Dam, people of the river basin took one of two positions, for or against; some sought quick construction, whereas others had a history of deploying oppositional movements. The complex nature of the grievances and hopes held by those living in the river basin suggests that we should avoid trying to make sense of the situation with a simplified notion of good and evil. The region's current problems cannot be explained by simple theoretical arguments of development critics, such as "discontinuation of the dam plan = a win against development." Instead, the fundamental problem remains to be confronted, even today.

This is a problem of consensus building around public development projects in hard-path style water resources management. Because the infrastructure supporting the basin is not something that can be built quickly, in order to implement a plan, it is necessary to obtain the consent of many stakeholders. In many cases, a considerable amount of time is required for the consensus-building process from planning to implementation, and a diversity of strategies and tactics among stakeholders, both within and without the basin, will need to be developed. Such coordination of advantages and disadvantages is generally not easy, and in the case of dam construction based on IWRM, there are cases in which construction has not started on a project area even after several decades have passed. However, in general, most development projects proceed as progress described in the explanatory diagrams: amidst long-standing interactive bargaining processes, those in opposition remain silent and allow consensus building to proceed. This phenomenon is seen in the Niu Dam construction plan area as well.

As such, the discontinuation of a dam project that was a symbol of modern IWRM, coupled with the simultaneous ushering in of a regional plan drawn up by the region that had just canceled the dam, invites a situation in which the futures of all people affected by the plan must be rewritten, and that is no empty theory. Rather, it is something that can raise awareness of the fact that "the general idea of integration in IWRM is in actuality not only about water resources but also includes regional planning." However, in today's debate around the Niu Dam, this point has been left behind: area residents thought about their plight, having been continuously toyed with by the authorities for many years, and decided it was necessary to make an effort to rebuild for the future.

Administering to the condition in Old Yogo Town from the perspective of water resources management, we need to confront two problems: (1) a period of infrastructural renewal for not only rural community drainage (rural sewers) but also irrigation drainage facilities established by the Kohoku General Development Project and the Lake Biwa General Development Project and (2) a plan to be decided for the region that has completed the resettlement and abandoned the dam, which is now in an advanced state of disrepair. Even though the dam project was canceled, area residents are trapped within this problem and have to think about planning for the future, all the while taking into consideration the major social problems of Japan: decreasing birthrate, aging population, and overall population decline. Moreover, they have to overcome the impact of past responsibilities and discord between people caused by more than 30 years of problems with the Niu Dam. In particular, in the upper part of the basin where the relocation of people is complete, the minimum land necessary for human habitation no longer remains, and even the roads running through the area are no longer in a useable condition (Interview; Mr. Nobuo Murakami). Furthermore, it is necessary to think about future users and future generations when considering updating various kinds of agricultural irrigation facilities: updating existing facilities in the present economic condition will place an enormous socioeconomic and environmental burden on the region. Therefore, it is important to think in terms of "positive development," which reduces the future load on the regional community.

However, planning for the future of actual regions is no easy task. The reason is that development in most municipalities is often based on plans drawn up by national or regional public groups, much like the integrated plans of the past.

Further, the reality is that the majority of regional planning in local governments is produced by municipalities and consultants cooperating together to coordinate regional stakeholders. For that reason, in a case like the discontinuation of the Niu Dam, where claims of "government failure" can certainly be made, we don't think the actual implementation of "future planning by local residents themselves" will be any easy task.

However, the fact is that national and regional public groups are unable to come to terms with the cancelation of the dam and waste the present moment on fool's errands. Lacking any sense of reevaluation or reconsideration, new development plans are being put forward that merely repeat the past. Moreover, if one looks at the entire water system of the region, it cannot be said that the shift toward integrated planning proposed as part of the Mother Lake 21 Plan (Lake Biwa Integrated Conservation Improvement Project), commenced in 2000, after the completion of the Lake Biwa Integrated Development Project in the late 1990s, is necessarily proceeding smoothly. Even looking at it from this angle, it scarcely needs to be pointed out that there are two flaws in this plan for the future. In other words, it appears that the reactionary movements, which have for so long relied on the big story of the integrated development plan principally based on strong capital, continue to be trifled with today. Further, as a region that was so misled by the government's error, those living there now must think and make a choice, and when they do it will be important that the residents cooperate with scientists and other "outsiders" because they are drawing a future together. On this point, Lang's notion of the "co-creation of science and society" (Lang et al. 2012) is of particular importance.

In development plans, it is necessary to adopt a long-term perspective; however, there is a need to try to adapt to a future that the society still cannot see clearly. In the case of Old Yogo Town, the issues of "responsibility to the river basin" and "the problems of the region" remain major problems. The first challenge is to "protect the quality of the Lake Biwa–Yodo River water system," and "drawing a future for regions beset by a declining and aging population" is the second. In the integrated approaches of IWRM of the past, it was acceptable to draw out only one major plan, but these plans were made in the 1980s, and the favorable socioeconomic conditions of the time had an influence. But today, it is clear that institutions like national and prefectural governments are finding it difficult to maintain their supportive function, and an adaptive approach to the future is needed, wherein "the region thinks and decides for itself." However, how might a perspective be derived that differs from the traditional integrationist view? Invoking Asit K. Biswas (2008), the authors discuss this point and clarify the remaining challenges in the final section of this chapter.

2.4 Conclusion

First, Yogo Lake and the Takatoki River in Old Yogo Town make up a special region called the "Source of the Yodo River"; it is, therefore, important to take a soft-path approach that makes use of existing infrastructure and considers the risk to the natural environment and future generations of society. Therefore, the challenge then becomes how to formulate a soft-path approach that curtails social costs.

Keller et al. (1998) pointed out the limitations of development by classifying water resources management development as the growth phase of river basin development, with all development occurring in three phases (growth → maintenance → expansion into surrounding areas). That suggests that we can clarify our problem with the following two points. First, in areas with a reduced potential for development, there needs to be an effective measurement of demand, and resource management must be carried out. Second, in management of river basins entering the mature phase of development, blind adherence to tradition will no longer work. Considering these suggestions, we can say that development, including agricultural irrigation, in the river basin around Yogo Town is already mature; therefore, it is time to "conserve the river basin" with effective management.

Second, quoting an overview of the water resources issues that were attempted to be incorporated into IWRM, as exemplified by Asit K. Biswas (2008), issues the region considered important can be summarized as in Table 2.1 (issues considered important are underlined). Though there have been many integrated development projects in the past, the focus of integration thus far has been "water resources," and

Table 2.1 Task clarification for Yogo area's water management

1	Objectives which are not mutually exclusive (economic efficiency, regional income redistribution, environmental quality, and social welfare)	20	Upstream and downstream issues and interests
2	Water supply and water demand	21	Interests of all different stakeholders
3	Surface water and groundwater	22	National, regional, and international issues
4	Water quantity and water quality	23	Water projects, programs, and policies
5	Water- and land-related issues	24	Policies of all different sectors that have water-related implications, both in terms of quantity and quality and also direct and indirect (sectors include agriculture, industry, energy, transportation, health, the environment, education, gender, etc.)
6	Different types of water uses: domestic, industrial, agricultural, navigational, recreational, environmental, and hydropower generation	25	Intrastate, interstate, and international rivers
7	Rivers, aquifers, estuaries, and coastal waters	26	Bottom-up and top-down approaches
8	Water, the environment, and ecosystems	27	Centralization and decentralization
9	Water supply and wastewater collection, treatment, and disposal	28	National, state, and municipal water activities

(continued)

Table 2.1 (continued)

10	Urban and rural water issues	29	National and international water policies
11	Irrigation and drainage	30	Timings of water release for municipal, hydropower, agricultural, navigational, recreational, and environmental water uses
12	Water and health	31	Climatic, physical, biological, human, and environmental impacts
13	Macro-, meso-, and micro-water projects and programs	32	All social groups, rich and poor
14	Water-related institutions at national, regional, municipal, and local levels	33	Beneficiaries of the projects and those who pay the costs
15	Public and private sectors	34	Service providers and beneficiaries
16	Government and NGOs	35	Present and future generations
17	Timing of water release from the reservoirs to meet domestic, industrial, agricultural, navigational, environmental, and hydropower generation needs	36	National needs and interests of donors
18	All legal and regulatory frameworks relating to water, not only from the water sector but also from other sectors that have direct implications on the water sector	37	Activities and interests of donors
19	All economic instruments that can be used for water management	38	Water pollution, air pollution, and solid waste disposal, especially in terms of their water linkages
		39	Various gender-related issues
		40	Present and future technologies
		41	Water development, regional development, and any number of formulations and combinations of the above

Source: Based on Asit K. Biswas (2008) table

development has focused primarily on drainage and irrigation in particular. Furthermore, various types of integrated water utilization developments have completed many of the appropriate steps, and a series of policies toward solving water quality problems such as those at Lake Biwa have been developed.

Third, instead of IWRM based on a modern-style water storage model, a post-modern IWRM is necessary to address the region's present condition and the future that must still be drawn. There are many theoretical issues supporting this opinion. By closely examining current and future generations, it is clear we are at the stage of thinking through water management maturity. In other words, taking water as a clue to the reevaluation of the "entire region," we believe it is advisable to think about adaptive policies for the future.

It has also been pointed out by Asit K. Biswas (2008), but worth noting again, that water resources management accompanying development itself contains an element of regional development. However, in most cases up to the present, water resources management and regional development have been treated as separate. Yet, if we reorganize what we call water resources to fit within a social context and take a look at the problems that emerge, it is not true that regional development and preservation seem to have no relationship to water resources. In actual public policy, from the resolution of water resources problems to environmental education, there have been examples that have covered the other diverse matters of policy as well. However, the era when a "big development story" can be relied on, as has traditionally been the case in integrated development, has already come to a close. It is for this reason that a paradigm shift from this traditional issue of "integrated management" toward "how to adapt to the future" is indispensable.

Ultimately, methods are not needed for rationalizing the hard-path approach; the establishment of a soft-path approach through the continuous accumulation of carefully deliberated, small actions is required. This challenge requires consideration of the following two points. First, as a "presentation of the dam removal model," we need to draw up an image of a basin's future that demonstrates that both the regional preservation and optimal conservation of the water system are compatible without a dam. In particular, it is necessary to have multiple problem-solving alternatives from the planning stage onward. Developers should not exclusively create this multiplicity of alternatives. Rather, they should result from collaboration between local government and community residents.

Second, when drawing up a soft-path approach for the region, the factors that become problematic are not only water and the surrounding components but also the creation of a platform that can accommodate various kinds of service industries that are dedicated to the preservation of the region, including tourism. For challenges like these, it is necessary that the people of the region think things through for themselves, while making use of various "outsiders." Importantly, this time involves the proactive participation of the local entities who are the "relevant parties." Even if the solution is entrusted to outside experts, the ones who can take responsibility for the eventual outcome live in the region as well as their local government. And that is why it is necessary that those who will ultimately live with the consequences should think for themselves, take action, and draft a plan of how they can make a regional culture of the form that can begin to use various systems and outside experts for problems in the region.

As an area of outstanding impact on human civilization, the reconstruction of water resources management, including irrigation and flood control, fulfills a role more important than any other. In order to break free from the hard-path approach of the modern reservoir model, we need a multitude of small actions from people living in the region as they engage in trial and error, consider alternatives, and draw a shared future along with a diversity of stakeholders. This is the essential requirement for the participatory approaches of today.

Acknowledgment We wish to thank Mr. Nobuo Murakami for the advice on methodological design and considering the Niu Dam Plan and Lake Yogo. He politely agreed to be interviewed and invited us to meet with him. The Research Institute for Humanity and Nature (RIHN) "Designing Local Frameworks for Integrated Water Resources Management" Project (C-09-Init) also supported this work.

References

Amory LB (1977) Soft energy paths: toward a durable peace. Ballinger Publishing Company, Cambridge, MA, 265p

Barnard CI (1938) The functions of the executive, vol xvi. Harvard University Press, Cambridge, 334p

Biswas AK (2004) Integrated water resources management: a reassessment. Water Int 29(2):248–256

Biswas AK (2008) Integrated water resources management: a reassessment. Water Resour Dev 24(1):5–22

Ehrlich E, Landy B (2005) Public works, public health: directions for America's infrastructure (CSIS Reports). Center for Strategic & International Studies, Washington, DC, 24p

Global Water Partnership (2000) Integrated water resources management. Global Water Partnership Technical Advisory Committee, Background Papers (Tac Background Papers). No. 4, 67p

Hamasaki H, Kato H, Nakagiri T (2012) Possibilities of approach integrating 'hardware' and 'software' for sustainable water resources management: the case of Bali, Indonesia. In: Proceedings of the 4th International Water Association Asia-Pacific Young Water Professionals conference, 7–10 December, 2012, National Museum of Emerging Science and Innovation "Miraikan," Tokyo, Japan

Keller J, Keller A, Davids G (1998) River basin development phases and implications of closure. J Appl Irrig Sci 33(2):145–163

Lang DJ et al (2012) Transdisciplinary research in sustainability science: practice, principles, and challenges. Sustain Sci 7(Supplement 1):25–43

Maass A et al (1962) Design of water-resource systems: new techniques for relating economic objectives, engineering analysis, and government planning. Harvard University Press, Cambridge, MA, 638p

Maksimović Č, Tejada-Guibert JA (2001) Frontiers in urban water management: deadlock or hope. IWA Pub, London, 416p

Milly PCD, Betancourt J, Falkenmark M, Hirsch RM, Kundzewicz ZW, Lettenmaier DP, Stouffer RJ (2008) Climate change. Stationarity is dead: whither water management? Science 319(5863):573–574

Mukherjia A, Faconb T, de Fraiturec C, Moldena D, Chartresa C (2012) Growing more food with less water: how can revitalizing Asia's irrigation help? Water Policy (IWA) 14(3):430–446

Nakagami K (2015) Water management related to sustainability and human well-being: beyond the IWRM from local water sustainability. In: 10th RIHN international symposium, session 1 examining local water management: cases from Indonesia, Turkey and Japan, 18 June, 2015, Kyoto, RIHN

Reuss M (2003) Is it time to resurrect the Harvard water program? J Water Resour Plan Manag 129(5):357–360

Snellen WB, Scherevel A (2004) IWRM: for sustainable use of water 50 years of international experience with the concept of integrated water management (Background document to the FAO/Netherlands conference on water for food and ecosystems). Ministry of Agriculture, Nature and Food Quality, Wageningen, 15p

Svendsen M, Trava J, Johnson SH (2000) A synthesis of benefits and second-generation problems. In: Groenfeldt D, Svendsen M (eds) Case studies in participatory irrigation management. World Bank Institute, Washington, DC, pp 139–157

Chapter 3
Environmental Assessment in Collaboration with Local Residents

Budi Indra Setiawan, Satyanto Krido Saptomo, Yudi Chadirin, Chusnul Arif, Rudiyanto, and I Wayan Budiasa

Abstract Environmental assessment is key in designing a local framework for integrated water resources management dealing with land-use and climate change. We involved local residents in acquiring environmental data to clarify significant land-use and climate changes in watershed and field scales. In the watershed scale, a series of daily climate data was collected from three automatic weather stations, each available in the upstream, midstream, and downstream areas of the Saba Watershed from 2007 to 2014. The annual rainfall pattern changed; it decreased in all the stations after 2010. The annual rainfall downstream was always lower than at the other two stations located at higher elevations. As a consequence, the downstream area had the earliest and the longest dry season compared to the other areas. In the field scale, climate conditions were assessed on the basis of intensive measurements using automatic weather stations and soil sensors in three representative locations. Based on the monitored data in the context of climate change, the optimal planting date during a year was October 15 (first season), February 13 (second season), and June 12 (third season). We estimated that the irrigation water requirement was 108, 283, and 751 mm, respectively. We recommend constructing a rainwater reservoir to store more irrigation water.

Keywords Climate change • Environmental condition • Local people • Saba Watershed • Water balance • Paddy fields

---

B.I. Setiawan (✉) • S.K. Saptomo • Y. Chadirin • C. Arif • Rudiyanto
Department of Civil and Environmental Engineering, Bogor Agricultural University, Bogor 16680, Indonesia
e-mail: budindra@ipb.ac.id

I.W. Budiasa
Faculty of Agriculture, Udayana University, Jalan Kampus Bukit Jimbaran, Jimbaran, Kuta Selatan, Kabupaten Badung, Bali 80361, Indonesia

K. Nakagami et al. (eds.), *Sustainable Water Management*,
DOI 10.1007/978-981-10-1204-4_3

3.1 Introduction

A continuing increase in water use leads to growing concerns over the availability and quality of water supplies. As water becomes scarcer, there is a growing need to find ways to produce sufficient food to feed the world's expanding population while using less water, safeguarding fragile environmental services, and without much opportunity to convert new agricultural lands. Then, it is important to improve the management of water and land resources, with the aim of underpinning food security and reducing poverty, while safeguarding vital environmental processes. Research focusing on water availability and access is vital, including adaptation to climate change, how water is used and how it can be used more productively, water quality and its relationship to health and the environment, and how societies govern their water resources.

The outcome of the research should improve agricultural water management, enhance food security, protect environmental health, and alleviate poverty, especially in developing countries. Problematic issues that need to be addressed, among others, are population and welfare increases, increased demand for food and other agricultural products, a decrease of arable land because of land-use competition, uncertainty of available water because of climate change, shifting of the cultivation calendar, and a decreased number of farmers. Research outcomes must have the answers to the following questions: Can traditional farmers produce more food and other agricultural products with less and less water? Can the stakeholders manage the available water in order to increase water productivity? How can research results be delivered to the stakeholders so that they use data and information to support their decisions dealing with water resources management?

This paper reports a part of collaboration research results between the Research Institute for Humanity and Nature, Bogor Agricultural University, and Udayana University from 2011 to 2015 conducted in the Saba Watershed of Bali Province, which has been facing the problem of water availability in recent years that might be caused by the occurrences of climate change. Water management in the Saba Watershed has a long historical background, which involved many aspects of nature and human life. Even though stakeholders with their local wisdom have so far been capable of managing daily water resources, it seems that difficulties have arisen when dealing with the decline of available water because of climate and land-use changes. Data and information about these related matters, including awareness and alert systems, are not yet available. This research will analyze how water resources and water availability are related to, or determined by, climate change and land use; determine the main driving forces behind land-use changes; predict consequences or future trends if these changes are not addressed in managing water resources; and, finally, summarize research findings for the stakeholders so they become more capable of dealing with water scarcity and its consequences.

The main purpose of this research was to enhance stakeholders' capability with climate, land-use, and socioeconomic changes when managing water resources, with special objectives to address the following: (1) clarify significant climate, land-use, and socioeconomic changes; (2) discover interrelations of climate, land-use, and socioeconomic changes with water resources, water availability, and water utilization; and (3) enhance water-use efficiency and water productivity.

The scope of this research was limited to a watershed with special focus on elaborating interrelations among the following aspects, variables, and units, which were as follows: (1) climate, focusing on temperature, rainfall, and evapotranspiration (ET); (2) land use, focusing on forests, uplands, and paddy fields; (3) socioeconomics, focusing on variables associated with the Human Development Index; (4) water resources, focusing on rainfall and river flows; (5) available water, focusing on water intentionally collected or tapped for supporting socioeconomic activity; (6) water use, focusing on water used in agriculture; (7) land productivity, focusing on a quantity of agricultural product per hectare; (8) water productivity, focusing on a volume of water to produce a quantity of agricultural product per hectare; and (9) labor productivity, focusing on an equivalent number of workers to produce a quantity of agricultural products per hectare.

In designing a local framework for integrated water resources management, it is compulsory to involve stakeholders from the initiation of planning through the final end of evaluation for further improvements. The stakeholders should know not only their demand on water for their daily socioeconomic activities but also the status of water resources in their environment. The status of water resources is dynamic, influenced naturally by climate and land use. Climate changes might be occurring in a watershed. Land-use conversions might be occurring in conjunction with socio-economic development in the countryside. Land conversion might accelerate when the existing farmlands cannot support basic survival needs. Without comprehensively knowing the climate and land-use changes and their severe impact on water resources, stakeholders would do business as usual, which in turn would create uncertainty. Stakeholders should, however, pay attention to enhancing land and water, as well as labor productivities, by applying any appropriate means, based on scientific findings. Those interrelationships could be found by interdisciplinary and transdisciplinary analysis. Those interrelationships would become important knowledge to design a local framework for integrated water resources management.

The expected results of this research include the following: (1) information regarding climate, land-use, and water resource variabilities; (2) validated hydrologic models and their simulation results; (3) models to analyze interrelation of land use, water resources, and the socioeconomy; and (4) improved knowledge and capability to properly manage water resources. Expected outcomes of this research topic also include enhancing the knowledge base of young scientists and an improved capacity of stakeholders. The anticipated impacts of this research include the following: (1) trendsetting an integrated water resources management plan and implementation and (2) tightening mutual relationships among researchers, stakeholders, and their institutions.

3.2 Conceptual Approaches

The authors underline a conceptual approach depicting an interrelationship of climate, land use, and socioeconomics linked to water resources, water availability, and water utilization. The concept might be explained in the following ways:

1. Climate and land use together determine the quantity and quality of water resources. Any changes in the climate and/or land use, to some extent, would significantly change the water resources. These changes could be detected by looking at historical data using proper methods of analysis. From the results of this analysis, it would be possible to forecast what would happen to the water resources in the future.
2. Water availability or available water is a part of a water resource that is available for supporting human activities. Another part of a water resource is a base flow that should be maintained in the water body for the sake of environmental quality. In this sense, base flow might be considered environmental flow. The base flow or environmental flow ought to be determined by looking at historical data regarding the minimum water discharge.
3. A part of the available water is used for domestic supply, industry, and agriculture. Another part of the water supply might return to the water bodies and/or the environment as seepage, deep percolation, and/or drainage. Socioeconomic conditions must play important roles in determining the proportion of water being used. In a developing country like Indonesia, water used for agriculture totals more than 80 %, which also implies a significant amount of water loss or poor efficiency in water use. That is why water use in agriculture needs special attention. Increased water-use efficiency means that more water would be available for domestic supply and industries.
4. Socioeconomic conditions are dynamic, influenced by many aspects including externality. In a developing country like Indonesia, socioeconomic development, which relies on resources, often results in land expansion, conversion of land use, or land-use changes. These changes would impact the water resources because of an accelerated surface flow and less infiltration. It is then of interest to know if there are any interrelations between socioeconomic and land-use changes and to find out which variables are involved in socioeconomic aspects.
5. Another important question is: which variables describe the socioeconomic status? These variables should be easily understandable by the stakeholders and should be acceptable to the outside society. We decided to use the Human Development Index and regional income as variables to indicate the socioeconomic condition in the studied areas. The government of Indonesia has used these variables as indicators in rating development plans and progress in all regions. These variables are periodically measured by the National Statistics Agency and published every year on its website.
6. As previously recognized, agriculture consumes an incomparable amount of water when analyzing the other two users, domestics and industries. As a consequence, a deeper investigation into the following questions is merited: How is

the available water being used? What is the merit of using the water? Is there any significant impact on the environment? Are there any possibilities to improve the use of water in order to be more efficient? It is then necessary to apprehend water use associated with plant growth, yield, and productivity, as well as greenhouse gas emissions.
7. The other two water users cannot be excluded from the analysis by any means because, even though they use less water, the economic productivity of using the water might be a multiple higher than in agriculture. It should also be recognized that there is an elevating trend of water demand in these two sectors elsewhere.

Based on those descriptions, finding an appropriate tool that integrates those variables into a system of analysis, which is executable and can be used to forecast their trends in the future, is imperative.

3.3 Description of the Saba Watershed

3.3.1 Geography and Population

The Saba Watershed is located in the northern part of the Bali Province within the geographical coordinates of 114° 55′ 13.08″ east to 115° 7′ 7.68″ east and from 8° 10′ 50.16″ south to 8° 20′ 5.64″ south (Fig. 3.1). The area of the watershed is approximately 140.19 km^2, with the highest altitude approximately 5000 m from the mean sea level. A part of the watershed belongs to the Buleleng Regency (78 %); the remainder belongs to the Tabanan Regency (22 %). In the Buleleng Regency, there are four districts located in the watershed, which are listed as follows: (1) the Banjar, (2) the Busungbiu, (3) the Seririt, and (4) the Sukasada. In the Tabanan Regency, there is only one district called the Pupuan District.

The Seririt District had the largest population, amounting to 78.8 thousand persons or occupying 26 %, with a density of 705 persons per km^2. Based on the densities and area of these districts in the watershed, the Banjar District has the most populated area, totaling 14 thousand persons and accounting for 31 % of the total population in the watershed, which totaled approximately 45.3 thousand persons.

The Banjar District had the highest annual rice demand. In terms of rice productivity, the Busungbiu District was highest. Rice productivity was larger than the national rice productivity, which was less than five ton/ha. From this perspective, it can be said that rice production is not a major problem in the Saba Watershed.

3.3.2 Land Use and Soil Types

In terms of land use, mixed farmland occupies the largest area (8290 ha), followed by forests (2018 ha), paddy fields (2828 ha), dry land (435 ha), settlement (121 ha), water (102 ha), bushes (49 ha), and others (176 ha). Based on our analysis from 2000,

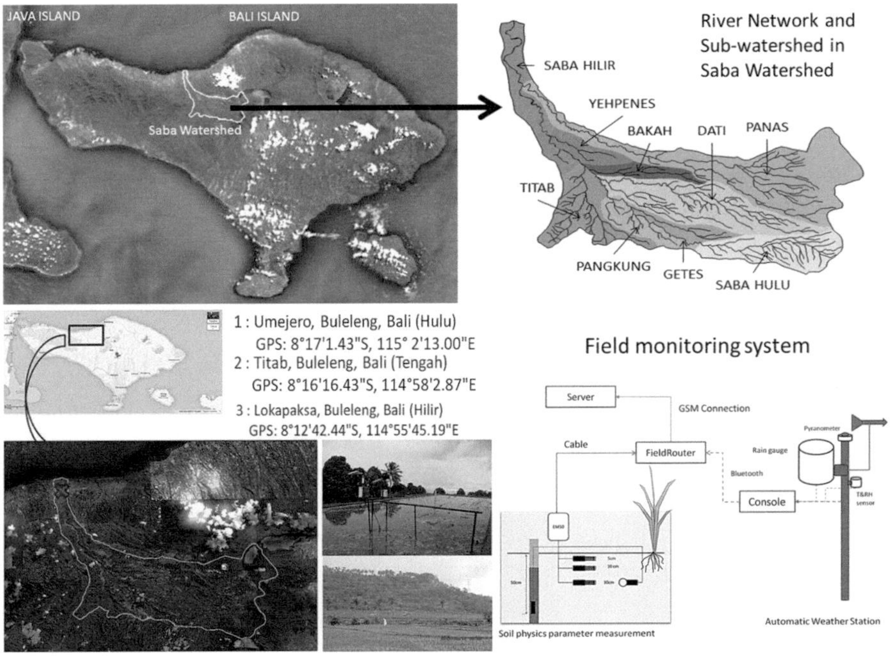

Fig. 3.1 Bali Island/Province and the location of Saba Watershed: river network, sub-watershed, and field monitoring system

land-use changes have been happening in which paddy fields increased by 17 %, whereas the area of mixed farmland and forest areas decreased by 13.2 % and 3.5 %, respectively. It is clear that the new paddy fields were converted from mixed farmland and forest. This means that more water is required to irrigate the ever-increasing area of paddy fields. There are five soil types distributed in the watershed: grayish brown andosol (29 %), reddish brown latosol (23 %), yellowish brown latosol (20 %), brown latosol (19 %), and gray regosol (9 %).

3.3.3 Rivers and Irrigation

There are nine sub-watersheds, which are Panas (32 %), Dati (19 %), Upstream Saba (11 %), Titab (11 %), Downstream Saba (9 %), Pangkung (5 %), Bakah (4 %), Yehpenes (4 %), and Getes (4 %). There are eight main rivers with the potential for irrigation, which are listed as follows: (1) the Saba River, (2) the Getas River, (3) the Jehe River, (4) the Selat River, (5) the Titab River, (6) the Bakah River, (7) the Panas River, and (8) the Ling River. There are also 28 dams and/or weirs. The existing irrigation network covers 9124 ha of paddy fields within 55 Subaks in 27 irrigation unit areas, including a newly developed Saba irrigation unit. The Saba irrigation unit area receives water from the Saba Dam covering 1915 ha, consisting of ten

Subak (five old Subak and five new Subak), supported by 25.25 km long secondary canals. In the Saba River, the Titab Dam has been constructed to supply additional irrigation water and to generate electricity.

3.3.4 Subak and Paddy Fields

Based on the UU 7/2004, the central government is responsible for the operation and maintenance of primary and secondary irrigation systems totaling more than 3000 ha, provinces for systems of sizes between 1000 and 3000 ha, and regency for irrigation systems up to 1000 ha. The Subak in Bali, however, remains fully responsible for the village- and the tertiary-level irrigation systems. Farmers may also request assistance from the government to upgrade tertiary irrigation and drainage facilities.

The Subak is a customary law with a socio-agrarian-religious nature that was established before the Dutch colonization. Furthermore, the Subak has developed as a landholding organization mainly related to water distribution for rice fields in one irrigation unit area that was legalized by Bali Provincial Decree No. 02/PD/DPRD/1972. The Subak plays a role in the operation and maintenance of the irrigation network, such as distributing irrigation water, conducting irrigation facility maintenance, managing conflicts, mobilizing/managing resources (land, water, human, and money), and conducting ceremonial activities.

The Subak consists of three main elements that are strictly interconnected and indivisible: (1) Parahyangan, (2) Pawongan, and (3) Palemahan. Daily life in the Subak societies is based on the philosophy of the Tri Hita Karana (three happiness causes). General and detailed rules for implementing it are included in the Subak Awig-Awig and Perarem (bylaws). The Subak system can be broadly subdivided into four major components: (1) the main structure (weir/inlet structure); (2) the main canal, with the function of conveying the irrigation water from the main structure upstream to the last rice field downstream; (3) the irrigation canals, with the function of distributing the irrigation water to the rice fields; and (4) the drainage facilities, including the small on-farm drains up to the big drains and rivers, serving several of the Subak irrigation systems.

The partition of irrigation water among Subak members is based on the principle of Ayahan. A portion of irrigation water, called Tektek, is the amount of water necessary for one season of irrigation of paddy fields with an area of up to approximately one hectare; other terms with the same meaning are Kecoran and Tanding.

Based on the Subak characteristics, Presidential Decree No. 3/1999, and the Water Law No. 7/2004 that is called as Participatory Irrigation Management (PIM), Subak serves multiple roles (e.g., to manage the irrigation system and conduct legal business units at the farm level). The successes of the Subak when conducting these multiple roles are very useful to support sustainable irrigated agriculture in Bali.

Based on cropping and the calendar, there are six cropping patterns associated with paddies and one pattern for other crops, such as vegetables. The typical and present cropping pattern and calendar in 2003 were used for the projection of the irrigation water demand. On average, rice productivity here is five ton/ha that is larger than the national figure. From this viewpoint, it is clear that rice production is not a major problem in the Saba Watershed because of the availability of water resources and the existing Subak in managing irrigation water.

3.4 Climate Conditions in Watershed Scale

From 2007 to 2014, a series of daily climate data was collected from three automatic weather stations available in the upstream, midstream, and downstream of the Saba Watershed. The data consisted of air temperature, rainfall, solar radiation, and evapotranspiration. Figure 3.2 shows that the annual rainfall varied with space and time. From 2007, the annual rainfall in the Munduk station decreased sharply with time; however, in the other two stations, the trends were positive. From 2010, however, the annual rainfall in all stations decreased with time. It seems that the highest

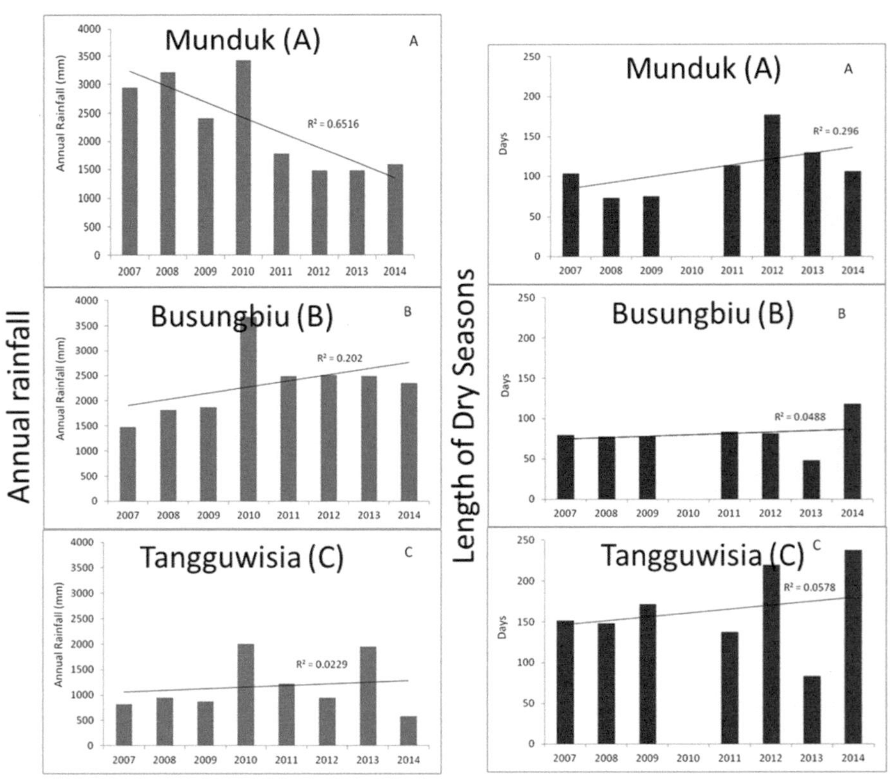

Fig. 3.2 Annual rainfall and length of dry seasons in Munduk (**a**), Busungbiu (**b**), and Tangguwisia (**c**) stations

rainfall in 2010 would not happen again in the future. As expected, the annual rainfall in the downstream station was always lower than the other two stations located in the higher elevations.

Figure 3.2 shows the start and length of the dry seasons in three areas of the Saba Watershed. In the upstream area, the earliest dry season occurred on April 29, 2012; its length was 117 days. The midstream area had the shortest and latest dry season. It started on June 7, 2012. As mentioned previously, the downstream area had the earliest and the longest dry season. The earliest dry season occurred in April 12, 2014, with a length of 237 days.

3.5 Climate Conditions in the Field Scale

Climate conditions in the field scale were assessed based on intensive measurements using automatic weather stations and soil sensors. These instruments were installed in three locations (Fig. 3.1) inside farmlands, each representing the upstream (the Umejero Village), midstream (the Titab Village), and downstream (the Lokapaksa Village). The farmlands belong to farmers so that they could directly participate in the observations and the management of these environmental monitoring systems. At least one local person attended in situ training to be capable of maintaining the instruments, to protect the instruments from intruders, to change batteries, and to deliver biweekly reports. Collaboration with local residents and related government offices has been established in the form of a stakeholder forum of the Saba Watershed.

Measured parameters included precipitation, solar radiation, potential evapotranspiration, air temperature, wind speed and direction, soil temperature, soil moisture, electrical conductivity, water table elevations, temperature, and matric potential. By applying a simple water balance equation, it is possible to evaluate water availability in the field scale and, furthermore, determine irrigation water requirements for plant cultivations.

Since 2013, three sets of field monitoring systems were installed in paddy fields, as well as an automatic weather station (AWS) in three different locations, which are upstream, midstream and downstream (Arif et al. 2014; Saptomo et al. 2014). The configuration can be seen in Fig. 3.1. Weather parameters were collected using AWS that consisted of sensors, including rain gauge, pyranometer, temperature and humidity sensors, and wind speed and direction sensors (anemometers). Meanwhile, soil physics parameters and water table elevations were measured using data loggers consisting of soil moisture, soil temperature, soil electrical conductivity, water depth, and soil matric potential sensors. There were three layers of soil measurement, 5 cm, 10 cm, and 30 cm of soil depth. All data were stored using Em50 data logger and the console for soil and weather measurements, respectively. Then, all data were collected by a field router and sent to the server daily, as well as field image data.

Although all environmental data were measured automatically using some devices and sensors, we collaborated with the local residents to maintain the system. Local residents checked the system regularly every week and recorded the information for the researchers. We trained them in how to maintain the system correctly and educated them regarding the importance of the data for optimizing adaptation and mitigation strategies in the face of climate change.

3.6 Water Balance in the Field Scale

Based on the field-monitored data, a simple water balance analysis was performed. There are two main components of a water balance, rainfall (as inflow) and evapotranspiration (as outflow). Evapotranspiration was estimated by using field-monitored data, such as solar radiation, air temperature, humidity, and wind speed data, according to the Penman-Monteith model (Allen et al. 1998).

In total, the difference between rainfall and evapotranspiration (ET) was positive, as shown in Fig. 3.3. It was indicated that inflow (rainfall) was higher than that of evapotranspiration. However, rainfall occurred in the specific time/event (discontinue),

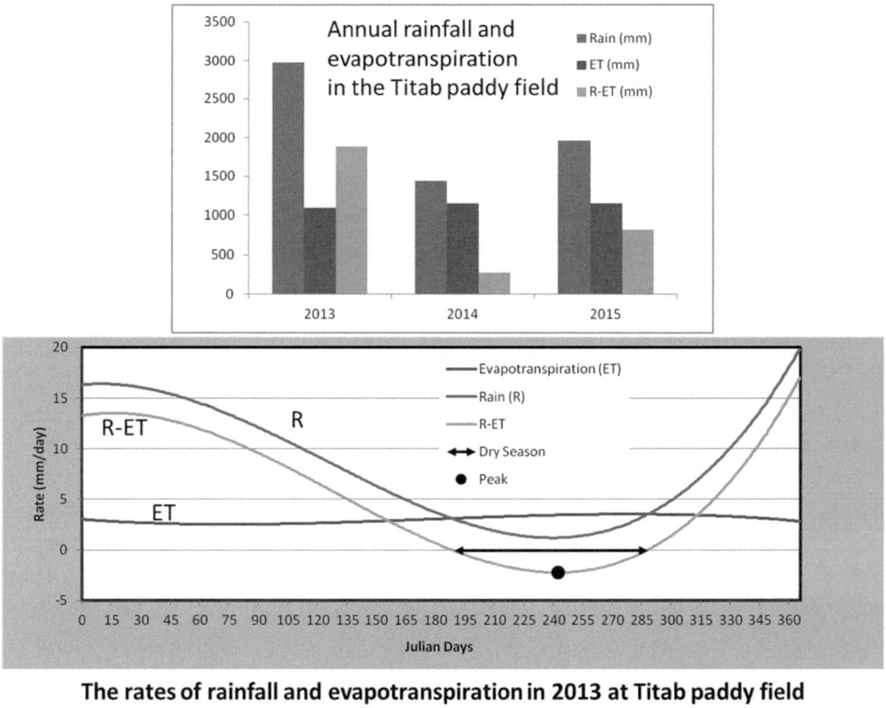

Fig. 3.3 Annual rainfall and rates of rainfall and evapotranspiration in the Titab paddy field (midstream)

and evapotranspiration occurred continuously, so it is important to analyze water balance daily.

Cumulative rainfall and evapotranspiration were fitted with the fourth-order polynomial equation (Setiawan 2015; Setiawan et al. 2015), and from the resulted equations, we can determine the onset and length of the dry season, as well as the rate of rainfall and evapotranspiration. Based on the rates, the irrigation water requirement can be determined. The irrigation water requirement is needed when the rate of evapotranspiration was higher than that of rainfall.

The midstream rates of rainfall and evapotranspiration in 2013 can be seen in Fig. 3.3. It was clear that during Julian days of 190 (July 8) and 289 (Oct. 15), the rate of rainfall was lower than that of evapotranspiration. It was indicated that the dry season and the irrigation water requirement are needed in this period. The peak of the dry season occurred when the difference between the rates of rainfall and evapotranspiration was at a minimum. It occurred on Julian days of 243 (August 30, 2013).

3.7 Drought Pattern

Regarding the drought pattern in the Titab paddy field during 2013 to 2015, there was a change in the dry season pattern, particularly in 2014. In 2013, the onset of dry season occurred on July 8 and ended October 15. The peak of dry season occurred on August 30. Then, in 2014, the pattern changed; the dry season arrived earlier, on May 27, and ended on October 28. In addition, the peak changed and came earlier, on August 24. In 2015, the longest dry season occurred with a total 154 days. Meanwhile, the earliest wet season started on October 15. This event can be used as guideline to determine planting dates and cultivation times during the year.

3.8 Irrigation Water Requirement

Based on the earliest wet season, the cultivating seasons and their water requirements for seasons I (wet season), II (wet-dry season), and III (dry season) are set forth in this section respectively. The planting dates (J-day) were established as follows: October 15 (289), February 13 (44), and June 12 (164). The harvesting dates were as follows: February 12, June 11, and October 10. The total rainfall (mm) was 726, 577, and 129 and ET 354 mm, 379 mm, and 400 mm. The water requirements in water depth (mm) were as follows: 108, 283, and 751. We estimated that each season was 120 days, with total rainfall at a minimum level (data in 2014), and total evapotranspiration was at the maximum level. We also assumed that the percolation rate was 4 mm/day, so total percolation during a season was 480 mm.

According to these estimations and assumptions, the total irrigation water requirements in season I, season II, and season III were 108, 283, and 751 mm, respectively. It is indicated that we should prepare more water in the dry season for irrigation. An alternative strategy to prepare water irrigation in the dry season would be to construct a rainwater reservoir.

3.9 Conclusion

Land-use and climate changes have affected water resources in the Saba Watershed. Suitable adaptation strategies are urgently needed, particularly for managing water in watershed and field scales. We performed an environmental assessment involving the local residents to acquire environmental data to clarify significant land-use and climate changes. The developed method was effective to determine the optimal planting date and irrigation water requirements in the context of climate change.

References

Allen RG, Pareira LS, Raes D, Smith M (1998) Crop evapotranspiration guidelines for computing crop water requirements. FAO – Food and Agriculture Organization of the United Nations, Rome

Arif C, Setiawan BI, Mizoguchi M, Saptomo SK, Sutoyo L, Budiasa IW, Kato H, Kubota J (2014) Performance of Quasi real-time paddy field monitoring systems in Indonesia. Pro Asia-Pacific Adv Netw 37:10–19

Saptomo SK, Setiawan BI, Arif C, Sutoyo L, Budiasa IW, Kato H, Nakagiri T, Kubota J (2014) Field monitoring stations network for supporting the development of integrated water resources management. Pro Asia-Pacific Adv Netw 37:30–41

Setiawan BI (2015) Analysis of drought pattern and water availability in rainfed agriculture. In: Proceeding of the workshop on strategic strengthening for south-south cooperation for modelling and managing hydro-hazards. UNESCO. Hotel Grand Kemang, Jakarta, 31 August–1 September 2015

Setiawan BI, Sutoyo, Budiasa IW, Kato H, Kubota J (2015) Seasonal variability of rainfall in Saba watershed of Bali Island. In: Proceeding of the fourth international workshop on CAAM-GRENE, Hanoi, 10–12 March 2015

Chapter 4
Local-Level Water Conservation Assessment in the Upstream Watershed Based on Land-Use Scenarios

Hiroki Oue and Sanz Grifrio Limin

Abstract To assess the effects of differences in land use in a mountainous sub-watershed on water conservation, namely, hydrological services like flood control and groundwater recharge, first, this chapter aims to present a water balance analysis in two sub-watersheds located upstream of the Saba River watershed. Specifically, the Titab and the Busungbiu-Tunju were compared. An annual water balance analysis in the two sub-watersheds in 2013 and 2014 revealed that the ratio of base flow to the total discharge was larger in the Titab, which has a lower areal percentage of clove plantation, a higher percentage of coffee plantation, and a slightly higher percentage of natural forest than the Busungbiu-Tunju. Second, by applying the International Center for Water Hazard and Risk Management/Public Works Research Institute (ICHARM/PWRI) distributed hydrological model, discharges under the present land use and three scenarios of changed land use were predicted. By converting all coffee plantations to clove plantations, base flow decreased, direct runoff increased, and the peak discharge increased. By converting clove plantations at high elevation to coffee plantations, base flow increased, direct runoff decreased, and the peak discharge decreased when compared with the present. In converting all land uses to natural forests, base flow was the largest, direct runoff was the smallest, and the peak discharge was the smallest of all cases. Comparison between the three land uses, coffee plantations, clove plantations, and natural forests, revealed that the clove plantation has the highest possibility of causing a flood disaster, the coffee plantation has a possibility of preventing a flood disaster and increasing groundwater, and the natural forest has the highest possibility of preventing a flood disaster and increasing groundwater.

H. Oue (✉)
Faculty of Agriculture, Ehime University, 3-5-7 Tarumi, Matsuyama 790-8566, Japan
e-mail: oue@agr.ehime-u.ac.jp

S.G. Limin
The United Graduate School of Agricultural Sciences, Ehime University,
3-5-7 Tarumi, Matsuyama 790-8566, Japan

K. Nakagami et al. (eds.), *Sustainable Water Management*,
DOI 10.1007/978-981-10-1204-4_4

Keywords Coffee plantation • Clove plantation • Natural forest • Water balance • Direct runoff • Base flow • Distributed model for discharge • Intake rate • Land-use scenario

4.1 Introduction

Natural tropical forests have been considerably converted among others to plantation forests in Indonesia. Though plantation forests are important for economic resources, they should be expected to have positive effects on hydrological services. Hydrological processes in some plantation forests, including coffee, have been studied widely; however, processes in a clove plantation forest have not been studied sufficiently.

In the upstream Saba River watershed in Bali, Indonesia, clove and coffee play an important role of being the main economic resources. Originally, these plantation forests were developed by converting natural forest area. From 1991 to 2014, 35.3 % of rice paddy fields in the upstream and midstream areas of this watershed were converted to clove plantations (Budiasa et al. 2014). This modern land-use conversion is due to the economic benefit of clove plantations over rice paddy fields and coffee plantations. Farmers' gross income from clove is Rp 450 million/ha with a yield of 0.3 t/ha, whereas coffee is Rp 11.25 million/ha with a yield of 0.75 t/ha. The gross income from rice is Rp 14 million/ha with a yield of 4.0 t/ha (personal communication). From the viewpoint of hydrological services, however, the role of each plantation forest in the Saba River watershed has not been well understood, although the local government has traditionally recommended coffee, claiming it would have more capability to intercept rainwater and would be better for water and soil conservation than other plantations or crops (Buleleng Regency Government 2010). However, the recommendation lacks supporting scientific evidence.

To clarify the roles of clove and coffee plantation forests on water conservation and compare discharges between the land uses, including natural forests in the upstream of the Saba river watershed, this study shows the present water balance in the upstream sub-watersheds, develops a hydrological model, and predicts changes in discharge between different land-use scenarios.

4.2 Targeted Region and Experimental Details

4.2.1 The Saba River Watershed and the Upstream Sub-watersheds

The Saba River watershed extends from 114° 56′ 33″ to 115° 06′ 30″ E and 8° 14′ 38″ S to 8° 20′ 14″ S, ranges in altitude from 0 to 2270 m, and encompasses an area of 152.8 km^2. The watershed belongs to two districts in the Bali Province, namely,

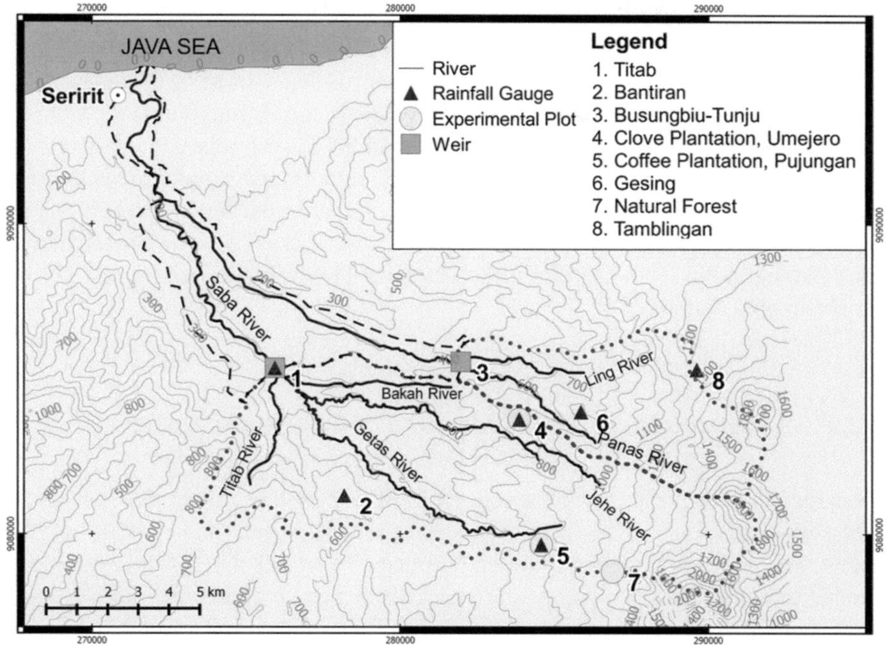

Fig. 4.1 Locations of (*1*) rice paddy field and the Titab sabo dam (205 m), (*2*) rice paddy field in Bantiran (610 m), (*3*) the Busungbiu-Tunju weir (405 m), (*4*) rice paddy field and clove plantation in Umejero (701 m), (*5*) coffee plantation in Pujungan (850 m), (*6*) residential area in Gesing (854 m), (*7*) natural forest in Pujungan (1460 m), and (*8*) flower garden in Tamblingan (1260 m) in the Saba river watershed. Two upstream sub-watersheds are surrounded by the dotted line

the Buleleng (78 %) and the Tabanan (22 %). In the Buleleng, four subdistricts are included in the watershed, the Banjar, the Busungbiu, the Seririt, and the Sukasada. By contrast, the Tabanan comprises only one subdistrict, the Pupuan. A topographical map of the Saba River watershed is shown in Fig. 4.1. In this watershed, plantation area accounts for about 59 % of all land uses. Rice paddy fields (both irrigated and rainfed) occupy 13 % and natural forest occupies about 12 %. Among the plantation areas, clove (*Syzygium aromaticum*) and robusta coffee (*Coffea canephora*) are two main commodities in the upstream area followed by rice in the downstream area.

This study focused on the upstream area of this watershed, aiming to clarify the roles of the two plantations on hydrological services, because the upstream area of the watershed must function to supply water resources to the downstream area and prevent a flood disaster in the downstream area. The upstream Saba River watershed has an area of 105.6 km^2 with elevation range 200–2270 m. The upstream watershed can be divided into two sub-watersheds: the Titab sub-watershed with outlet at the Titab sabo weir, No. 1 in Fig. 4.1, and the Busungbiu-Tunju sub-watershed with outlet at the Busungbiu-Tunju weir, No. 3 in Fig. 4.1. The Titab sub-watershed has an area of 75.6 km^2, and the Busungbiu-Tunju sub-watershed has an area of 30.0 km^2. The average elevation is 746 m in the Titab and 1020 m in the Busungbiu-Tunju.

In regard to the land use, there are differences between the two plantations. Both sub-watersheds are occupied mainly by clove plantations, but the Busungbiu-Tunju has a higher percentage (74.4 %) than the Titab (64.3 %). In turn, the Titab has a higher percentage of coffee plantations (18.3 %) than the Busungbiu-Tunju (3.8 %). One reason for the higher percentage of coffee in the Titab is a larger areal percentage at higher elevation, as coffee can be planted in highland. Another reason is the recommendation of coffee from the local government in the Titab. In addition, the Busungbiu-Tunju has a garden, mainly planted by flowers (7.7 %), whereas there is not a garden in the Titab. Natural forest occupies 11.6 % in the Titab and 10.6 % in the Busungbiu-Tunju.

4.2.2 Meteorological and Hydrological Measurements

From October 2012, the Davis Vantage Pro2TM was used to collect meteorological data (solar radiation, air temperature, humidity, wind speed, and rainfall) at four points. The points include the following, as depicted in Fig. 4.1: a flower garden in Tamblingan (No. 8), a rice paddy field adjacent to a clove plantation (No. 4), a rice paddy field (No. 2), and a rice paddy field near Titab sabo dam (No. 1). The logging interval is 10 min at Nos. 8 and 2 and 30 min at Nos. 4 and 1. Additionally, rainfall was measured in 10 min intervals by ECRN-100 Decagon at a coffee plantation (No. 5) since February 2013 and a residential area (No. 6) since November 2013. Meteorological data, except for rainfall, were applied to assist with the calculation of Penman's potential evaporation. Evapotranspiration was estimated by a water balance equation.

Discharge was measured at the outlet point of each sub-watershed, the Titab sabo weir and the Busungbiu-Tunju dam. Discharge was estimated by measuring the water level and converting the water level to the discharge by applying Q-H curve relation based on manual measurement. The water level was measured by placing a HOBO U20 water pressure sensor in each outlet. Manual measurement of discharge was conducted with a Kenek VE20/VET-200-10P electromagnetic current meter.

To assess the roles of different land uses on the hydrological processes, focus was placed on the infiltration process in addition to discharge. The intake rate was measured in the experimental clove (No. 4 in Fig. 4.1) and coffee plantations (No. 5 in Fig. 4.1) and in the experimental natural forest (No. 7 in Fig. 4.1). A PVC tube, the diameter of which was 11.4 cm, was installed into the ground at 10 cm or deeper and filled with water. Around the tube, a temporary concentric soil dike, the diameter of which was about 50 cm, was filled with water at the same time and with mostly the same depth as in the tube. In 2014, measurements were taken from the clove site on June 18 and 23, the coffee site on June 13 and 23, and the natural forest site on June 20 and 23, 2014. All measurements were performed on days when the antecedent precipitation was less than 2.0 mm for longer than 4 days.

4.3 Hydrological Traits in the Two Sub-watersheds in the Upstream Saba River Watershed

4.3.1 Water Balance in the Two Upstream Sub-watersheds

In 2013 and 2014, to compare the hydrological traits in the two upstream sub-watersheds, each annual water balance was estimated. The areal monthly rainfall was estimated by applying the elevation regression, that is, the relationship between the elevation and rainfall in each month, which was developed with measured rainfall at points in each watershed. Measured discharge was divided in direct runoff, which consists mainly of flood component and base flow, which consists mainly of groundwater runoff, by simply setting a threshold discharge between the two. The threshold discharge was considered from the recession process after rainfall in hydrographs: 0.188 mm/day in the Titab and 0.159 mm/day in the Busungbiu-Tunju. The evapotranspiration ratio, (ET/Ep), where ET is evapotranspiration estimated by water balance and Ep is Penman's potential evaporation, was approximately 1.0. In some types of forest under different conditions of wetness and vegetation, Takase et al. (1998) presented that the ET/Ep ranged from around 0.3 to 1.3 as in the relationship between annual rainfall and ET/Ep. They showed around 1.0 or higher ET/Ep in case of annual rainfall over 2000 mm/year. As a whole, ET/Ep and, thus, water balance in this study can be evaluated to be reasonable. ET and ET/Ep were larger in the Titab, which is likely caused by higher air temperature in higher elevations, in addition to the different land uses compared with that in the Busungbiu-Tunju. Thus, the share of available water resources, i.e., the total discharge, was smaller in the Titab. But, the ratio of the base flow to the total discharge was larger in the Titab (70.3 %) than in the Busungbiu-Tunju (60.5 %).

4.3.2 Comparison of Intake Rates Between Land Uses: Coffee Plantation, Clove Plantation, and Natural Forest

A larger direct runoff has a higher possibility to cause flood, whereas a larger base flow has a higher possibility to serve a stable water supply. The difference in the partitioning of discharge into direct runoff and base flow between the two sub-watersheds would be mainly caused by the difference of land use; there is a higher percentage of coffee in the Titab and a higher percentage of clove in the Busungbiu-Tunju. To assess roles of different land uses on the hydrological processes in a plot scale, an intake rate was compared between the two land uses and the natural forest.

An intake rate at 60 min after the experiment started was 12.0, 81.9, and 111.0 (mm/h) in the experimental clove (No. 4 in Fig. 4.1), coffee (No. 5 in Fig. 4.1), and natural forest (No. 7 in Fig. 4.1) sites, respectively. The rate at the natural forest site was at the same level as the final infiltration, 127 mm/h in a natural forest in East

Kalimantan Muhammad Farid (1996), and a little smaller than 220 mm/h at 60 min in a lowland natural forest in Central Kalimantan (Suryatmojo 2014). The rate at the coffee site could be the same level as in these natural forests, though it was a little smaller. The rate at the clove site, however, was about one order smaller than those in the natural forest site and the coffee site. One of the main reasons of the smaller infiltration rate in the clove site may be because fallen leaves are collected for extracting clove oil from them. Clove plantation management would be a disadvantage for forming permeable organic soil layer and retaining water in the ground. In turn, the coffee site is thought to have formed a permeable surface soil layer because of the fallen leaves of coffee and shade trees, similar to a natural forest.

The larger intake rate in the coffee plot may be one of the main reasons for a higher ratio of base flow and a lower ratio of direct runoff to the total discharge in the Titab. In terms of total land area, the Titab has a higher percentage of coffee plantations and lower percentage of clove plantations than the Busungbiu-Tunju. This hydrological property in the surface soil layer likely caused a smaller threshold discharge between direct runoff and base flow in the Titab as in Sect. 4.3.1, which was applied from the recession process in measured hydrographs.

In Sect. 4.4.3, a hydrological model of the Titab sub-watershed in 2013 is presented. The Titab was selected because the coffee plantation is still prevailing, compared with the Busungbiu-Tunju, and the effects of both increasing and decreasing coffee plantation areas could be preferably discussed. Then, in Sect. 4.5, by applying the model in the Titab under different land-use scenarios, the effects of a land-use change on discharge will be discussed.

4.4 Application of a Distributed Model for Estimating Discharge in the Titab Sub-watershed

4.4.1 Basic Concept of the ICHARM/PWRI Distributed Hydrological Model

The International Center for Water Hazard and Risk Management (ICHARM) and the Public Works Research Institute (PWRI) developed a distributed model for flood analysis. The ICHARM/PWRI distributed hydrological model is a runoff analysis model converting rainfall into runoff for a given watershed that can be classified as both conceptual and parametric and is a physically based, fully distributed model.

4.4.2 Spatial Data Preparation

Spatial data required for applying the model in this study include a digital elevation model (DEM), a land-use map, and an areal rainfall map. DEM and reference land-use maps can be obtained freely from USGS, i.e., GTOPO30 data and Global Land Cover Characterization (GLCC), respectively.

Data obtained from GTOPO30 have approximately 1 km resolution. Because of the relatively small area of the Titab sub-watershed, resolution of the grid for the model should be less than 1×1 km. In this model application, GTOPO30 data of approximately 1 km resolution were interpolated for the resolution of 0.5×0.5 km.

GLCC data are based on 1 km Advanced Very High Resolution Radiometer (AVHRR) satellites data, which spanned from April 1992 to March 1993. To revise present land-use conditions in 2013, raster data from GLCC were edited by referencing the present land-use condition, which was derived from Google Earth satellite images obtained on October 10, 2012 and July 29, 2013. The land use was interpreted manually by looking at the shape of the land use from the satellite images and by collating it with our ground truth. Types of land use that can be interpreted were clove plantations, coffee plantations, natural forests, water bodies, paddy fields, upland fields/gardens, and residential areas. Because each land use had a unique shape when viewed from the sky, it could be distinguished manually from the satellite image. Land uses in the upstream Saba River watershed were digitized and then converted to 1×1 km raster format of the GLCC database. In the land-use classification with GLCC database, this study took evergreen broadleaf forest (code, 421) for natural forest, mixed forest (code, 430) for coffee plantations, cropland/woodland mosaic (code, 290) for clove plantations, herbaceous wetland (code, 620) for rice paddy fields, and urban/built-up land (code, 100) for residential areas. Modified 1 km land-use maps were interpolated for higher-resolution application, i.e., 0.5×0.5 km. A higher resolution than 0.5 km was avoided to reduce the calculation time for the ICHARM/PWRI model.

An areal rainfall map was developed by applying an elevation regression, i.e., a relationship between the elevation and rainfall, with measured rainfall at the four measurement points in the Titab sub-watershed (Nos. 1, 2, 4, and 5 in Fig. 4.1). Default settings for areal rainfall in the ICHARM/PWRI model were determined by the Thiessen polygon method. When applying the elevation regression into this model, an hourly areal rainfall map was edited by applying steps as follows:

1. Creating a topographic map with 200 m intervals
2. Dividing intervals into several polygons
3. Determining the centroid point for each polygon
4. Applying the centroid points as points of dummy rain gauges
5. Applying elevation regression to fill rain data in each dummy rain gauge point
6. Exporting the dummy rain data to the ICHARM/PWRI model for developing hourly areal rainfall maps

As rainfall is the main input in the ICHARM/PWRI model, areal rainfall maps should be developed for each time interval for calculation. A possible shortest calculation interval in the ICHARM/PWRI model is 60 min, which means there should be 8760 areal rainfall maps developed for a year calculation. A total of 45 dummy rain gauge points were established in the Titab. Hourly rainfall at each dummy rain gauge point was calculated by an elevation regression. The relation was decided in each rain event with the measured rainfall at the four observation sites.

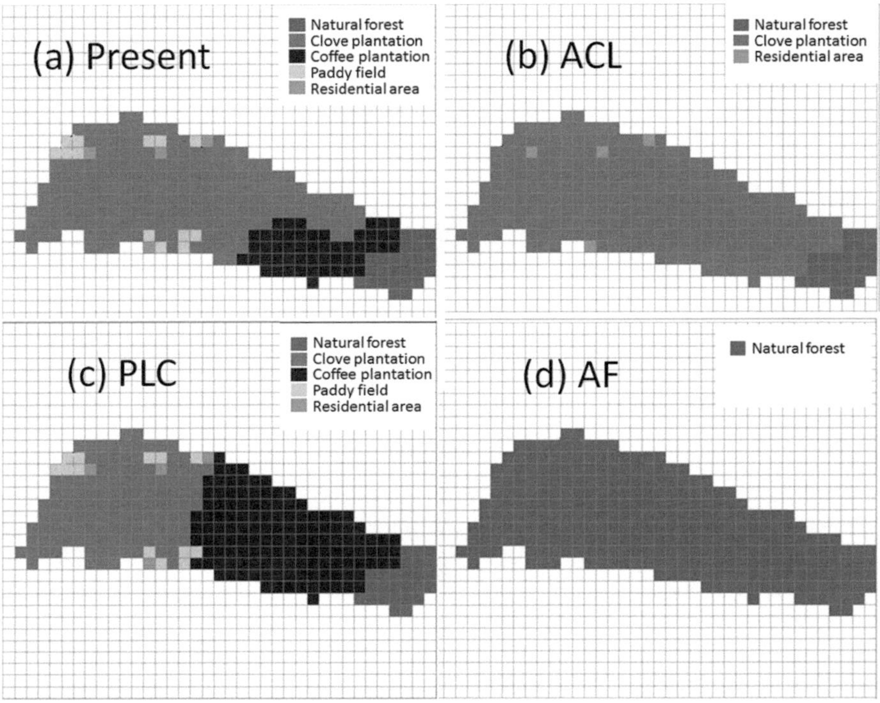

Fig. 4.2 (**a**) The present land use and land-use change scenarios; (**b**) ACL, (**c**) PLC, and (**d**) AF in the Titab sub-watershed applied to the ICHARM/PWRI model with 0.5 × 0.5 km grid

4.4.3 Parameter for the Model in the Titab Sub-watershed

The Titab sub-watershed was divided into 0.5 × 0.5 km grids, the present land uses of which are shown in Fig. 4.2a, and the parameters were attached for each grid. Untuned parameters were set for each land use. Parameters in the surface tank and the aquifer tank were tuned by a trial-and-error method until the calculated discharge would be close enough to the measured discharge. In this process, parameters of each land use in the aquifer tank were set to be similar to each other because of the lack of information about the geology in the sub-watershed. The river tank and the river course parameters were based on default values derived from DEM.

The tuning process started from a default value of the GLCC/USGS database, the land-use classifications of which were written in Sect. 4.4.2. The comparison between measured and modeled hydrographs is shown in Fig. 4.3a. The total annual rainfall in this sub-watershed was 3082.0 mm/year in 2013. The total annual discharge calculated by the model was 1594.2 mm/year, which consisted of base flow (1013.5 mm/year) and direct runoff (580.7 mm/year), whereas the measurement was 1543.8 mm/year, which consisted of base flow (1130.0 mm/year) and direct runoff (413.7 mm/year).

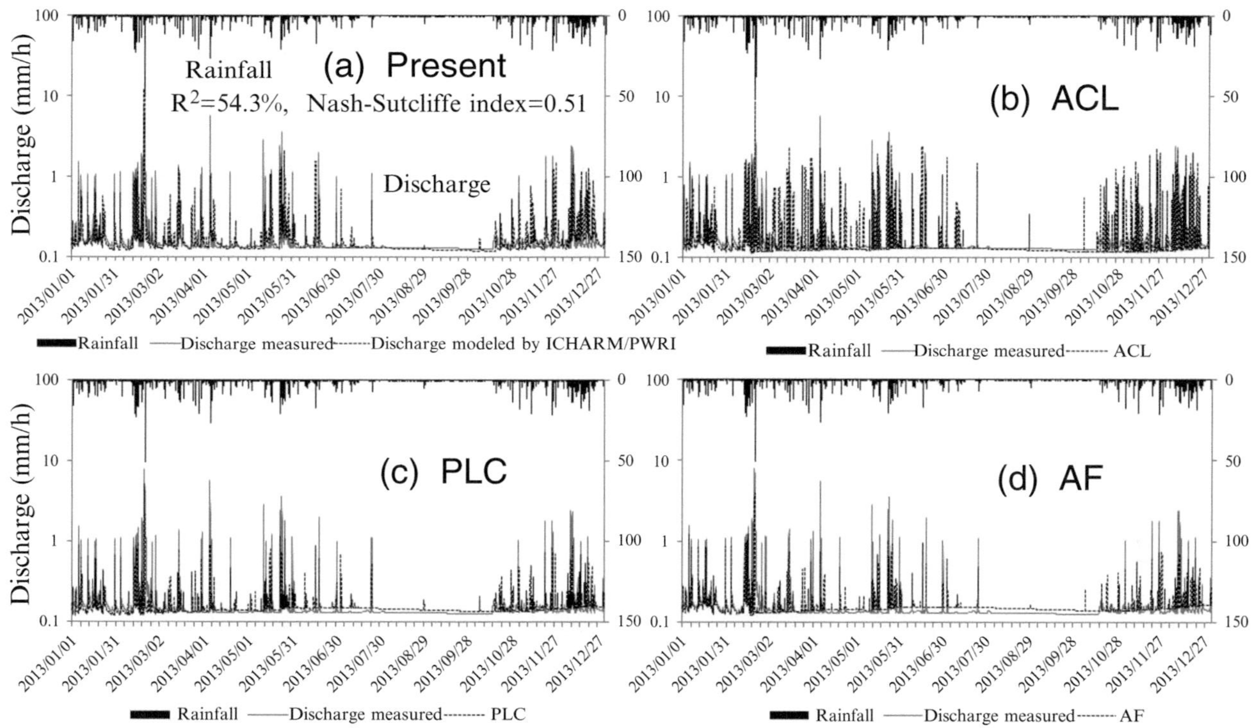

Fig. 4.3 Measured and modeled hydrographs (**a**) under the present land use and land-use change scenarios; (**b**) ACL, (**c**) PLC, and (**d**) AF applied to the ICHARM/PWRI model in the Titab sub-watershed in 2013

The determination coefficient for the fitted hydrograph was 0.543 and the Nash-Sutcliffe index was 0.51. With these, the applicability of the model could be validated. If the application results in the ICHARM/PWRI distributed model were compared with those in a lumped tank model, though the details of a lumped tank model are not shown here, the determination coefficient was higher in the distributed model than in a lumped tank model (0.229), and the Nash-Sutcliffe index was closer to 1.0 in the distributed model than in a lumped tank model (−0.99). Advantages are in the distributed model, which can consider effects of topography, i.e., distributions and sequences of land use and the slope on the water flow to the river channel. By contrast, a lumped tank model considers just the land-use effect by taking the ratio of each land use to the total area.

4.5 Application of the ICHARM/PWRI Distributed Model for Land-Use Change Scenarios

Three land-use scenarios applied are as follows:

1. The conversion of all coffee plantations to clove plantations (ACL) as shown in Fig. 4.2b. This scenario is applied by replacing all coffee plantations and paddy fields with clove plantations. This scenario corresponds to the trend of land-use change in this sub-watershed, where farmers prefer clove plantations because of higher commercial value of clove.
2. The conversion of clove plantations, located at the elevation higher than 500 m, to coffee plantations (PLC) as shown in Fig. 4.2c. This scenario corresponds to the previous land-use condition before coffee had been replaced by clove in 1970s. According to interviews with elder farmers on-site, coffee was the main commodity in the area at higher altitude in this sub-watershed before 1970s.
3. The conversion of all land uses to natural forest (AF) as shown in Fig. 4.2d. This scenario attempts to represent the natural condition before all the plantations, agricultural fields, and residential areas had been established.

All the scenarios were applied without changing the present natural forest area. Present natural forest in this sub-watershed is a conservation area under the management of the Balai Konservasi Sumber Daya Alam (BKSDA) Bali, i.e., the Natural Resource Conservation Agency Bali, Ministry of Forestry.

The predicted discharge by the model with ACL land-use scenario is shown in Fig. 4.3b. The predicted total annual discharge was 1733.1 mm/year, which was larger than the present total discharge (1543.8 mm/year). The predicted discharge consisted of base flow (943.5 mm/year) and direct runoff (789.6 mm/year). Thus, it was predicted that base flow would decrease and direct runoff would increase under this scenario. The predicted maximum daily discharge (12.1 mm/day) was larger than that in the present (8.0 mm/day). These changes were likely caused by the properties of a clove plantation, which could produce larger surface flow and retain less water in the ground compared with the coffee plantation. The conversion of all coffee plantations to clove plantations would increase the flood disaster risk in this sub-watershed.

The predicted discharge by the model with PLC land-use scenario is shown in Fig. 4.3c. The predicted total annual discharge was 1458.4 mm/year, which was smaller than the present. The predicted discharge consisted of base flow (1147.0 mm/year) and direct runoff (311.4 mm/year). Thus, it was predicted that base flow would increase and direct runoff would decrease under this scenario. The predicted maximum daily discharge (4.6 mm/day) was smaller than the present. These changes likely have been caused by properties of a coffee plantation, which could retain more water in the ground. The retained water would have increased the base flow from the aquifer layer. The conversion of clove plantations to coffee plantations would decrease the flood risk and increase the availability of groundwater or river water, especially in a dry period.

The predicted discharge by the model with AF land-use scenario is shown in Fig. 4.3d. The predicted total annual discharge was 1423.0 mm/year, which was smaller than the present. The predicted discharge consisted of base flow (1181.9 mm/year) and direct runoff (241.1 mm/year). Thus, it was predicted that base flow would increase and direct runoff would decrease even more than in PLC. The predicted maximum daily discharge (4.0 mm/day) was the smallest of all cases. The conversion of all land uses to natural forest would have a slightly higher advantage than the PLC scenario in decreasing the flood risk and increasing base flow, especially in a dry season.

4.6 Conclusion

To assess the effects of differences in land use on hydrological processes, this study firstly compared the water balances in two sub-watersheds with different land uses, the Titab and the Busungbiu-Tunju. Both sub-watersheds are mainly occupied by clove plantations but with a higher areal percentage in the Busungbiu-Tunju than in the Titab. The Titab has a higher areal percentage of coffee plantations and a slightly higher percentage of natural forest than the Busungbiu-Tunju. An annual water balance analysis in the two sub-watersheds performed in 2013 and 2014 revealed that the ratio of base flow to the total discharge was larger in the Titab (70.3 %) than in the Busungbiu-Tunju (60.5 %). The difference in the partitioning of discharge into the two components was likely caused by different land uses in the two sub-watersheds.

Secondly, the intake rate was compared between the two land uses and the natural forest to assess roles of different land uses on hydrological processes in each experimental site. The rate at the coffee site was a little lower than or at the same level as that at the natural forest site or natural forests of the previous studies. However, the rate at the clove site was about one order smaller than those of the natural forest and the coffee sites. The lesser infiltration of the clove was likely caused by collecting fallen leaves for extracting clove oil from them. Such a management of the clove plantation was found to be a disadvantage for forming a permeable organic soil layer and retaining water in the ground. The coffee plantation was evaluated to form a permeable surface soil layer because of the fallen leaves of coffee trees and also of shade trees, similar to a natural forest. As a result, the larger intake rate in the coffee plot was considered to cause the higher ratio of base flow and the lower ratio of direct runoff to the total discharge in the Titab.

Thirdly, to assess the effects of differences in the land use in the Titab subwatershed on water conservation, namely, hydrological services such as flood control and groundwater recharge in the total watershed, this study applied the ICHARM/PWRI distributed hydrological model for reproducing discharge under the present land use and some scenarios of changed land uses. The applicability of the model could be validated by 0.543 of the determination coefficient and 0.51 of the Nash-Sutcliffe index in fitting the measured hydrograph.

Lastly, three land-use scenarios were input to the model so that the effects of changes in land use on water conservation could be assessed. The scenarios are the conversion of all coffee plantations to clove plantations (ACL), the conversion of clove plantations located at the elevation higher than 500 m to coffee plantations (PLC), and the conversion of all land uses to natural forest (AF). In the ACL, the base flow decreased, the direct runoff increased, and the peak discharge increased compared with the present. In the PLC, the base flow increased, the direct runoff decreased, and the peak discharge decreased compared with the present. In the AF, the base flow increased and direct runoff decreased, even more than in PLC. The peak discharge was the smallest in the AF of all cases. Comparing the three land uses, coffee plantations, clove plantations, and natural forest, it was predicted that the clove plantation has the highest possibility to cause a flood disaster, the coffee plantation has a possibility to prevent a flood disaster and increase groundwater, and the natural forest has the highest possibility to prevent a flood disaster and increase groundwater. From the perspective of both hydrological services and economic benefit in the Saba River watershed, the PLC scenario would give highest benefit of all cases, including the present situation.

References

Budiasa IW et al (2014) The role of Subak facing land use changes and tourism development within Saba watershed, northern of Bali province Indonesia. In: Proceedings of ISSAAS 20th founding anniversary celebration & 2014 international congress and general meeting, Tokyo, Japan, 8–10 November

Buleleng Regency Government (2010) Laporan status lingkungan hidup daerah Kabupaten Buleleng (Environmental condition report of Buleleng Regency), Singaraja; 33 (in Indonesian)

Muhammad Farid AR (1996) Site feasibility study for peat swamp forest in peninsular Malaysia. Report for FJ Resource Teck. SDN. BHD

Suryatmojo H (2014) Recovery of forest soil disturbance in the intensive forest management system. Procedia Environ Sci 20:832–840

Takase K et al (1998) Properties of annual evapotranspiration from the catchments in a semi-arid zone and in the south western part of Japan. J Jpn Soc Hydrol Water Resour 11(7):694–701 (in Japanese)

Part III
Practices in Indonesia: Participatory Approach Toward Sustainable Water Resources Management

Chapter 5
A Participatory Approach to Enhance Multistakeholders' Participation in the Saba River Basin

I Wayan Budiasa and Hisaaki Kato

Abstract Integrated water resources management (IWRM) for the Saba River Basin began as a pilot project undertaken by the Research Institute for Humanity and Nature (RIHN), Japan and Bogor Agricultural University (IPB), Indonesia (2012–2016). A participatory approach through multistages of multistakeholders' meetings has enhanced participation in implementing the IWRM concept. The research project has successfully improved the multistakeholders' capacity with regard to IWRM; their participation in strength, weakness, opportunity, and threat (SWOT); and identification of key development issues within the Saba River Basin. They have also improved the decision-making processes through every stage as identified by the accomplishment of some important common agreements, such as the need for a forum for IWRM entitled the Saba River Basin Community (the "Community"), as well as its vision, role, function, organization structure, basic rules, priority programs from 2015 to 2020, and action plans in 2015/2016. However, the Committee must now transition into a concrete activity and push for action through a process of trial and error that will build the future. Obviously, the Community will be more effective in fulfilling its roles if it has appropriate financial and legal support.

Keywords Participatory approach • Water management • Multistakeholders' meeting • Community establishment • Action plan • Saba River

---

I.W. Budiasa (✉)
Faculty of Agriculture, Udayana University, Jalan Kampus Bukit Jimbaran, Jimbaran, Kuta Selatan, Kabupaten Badung, Bali 80361, Indonesia
e-mail: wba.agr@unud.ac.id

H. Kato
Research Institute for Humanity and Nature,
457-4 Motoyama, Kamigamo, Kita-ku, Kyoto 603-8047, Japan

K. Nakagami et al. (eds.), *Sustainable Water Management*,
DOI 10.1007/978-981-10-1204-4_5

5.1 Introduction

Water issues affect all sectors of society and economy. Population growth, rapid urbanization and industrialization, the expansion of agriculture and tourism, and climate change put water under increasing stress (GWP and INBO 2009). The issue of integrated water resources management within the river basin started with the UN Mar del Plata Conference in 1977, which called for comprehensive development and planning in river basins. In 1977, Okun, a renowned American environmental engineer, promoted the concept of "integrated" water management in his published book, which became an example for US river basins. A comprehensive river basin authority was established in 1974 for the Citarum River, Indonesia (Jatiluhur Authority Corporation), styled after the Tennessee Valley Authority. Until the 1970s, the water resources in the basin were still considered to be augmented merely through infrastructure development and technical planning (Alaerts and Muigne 2003). Four Dublin principles in IWRM were promoted in the 1992 UN International Conference on Water and the Environment that recognize the interactions between water, land, the users, the environment, and the infrastructure that stores and guides the water. One of the principles states that water development and management should be based on a participatory approach involving users, planners, and policy makers at all levels. Furthermore, the Global Water Partnership (GWP) defined IWRM as "a process which promotes the coordinated development and management of water, land and related resources, in order to maximize the resultant economic and social welfare in an equitable manner without compromising the sustainability of vital ecosystems" (Hassing et al. 2009).

The river basin has been recognized as a practical hydrological unit for water resources management (GWP and INBO 2009). The Saba River Basin is one of 391 river basins in Bali-Penida. It covers an area of approximately 132.89 km^2 (BWS Bali Penida 2011) spanning the Buleleng and Tabanan Regencies (BPDAS Unda-Anyar 2008). It includes irrigated paddy fields totaling approximately 4156.82 ha, which are supported by eight main rivers (the Saba, the Getas, the Dati, the Jehe, the Bakah, the Titab, the Panas, and the dan Ling) (BWS Bali-Penida 2005), and is managed by 56 Subak systems (Budiasa et al. 2015). Its water flows are usually up to four times higher during the wet season compared to the dry season, and the water quality is still suitable for rice culture and public water demands (Nakagiri et al. 2013); however, it is not suitable for drinking water, especially downstream.

The Saba River Basin has many development issues involving soil erosion hazards, land cover, agricultural land use changes, water degradation, competitive use of water resources, and inefficient irrigation channels. Heavy and very heavy soil erosion hazard levels seemly occurred because of land cover changes in the upstream and in central areas of the basin. Heavy soil erosion hazards (180–480 t/ha/year) have been indicated at the Pujungan, Pupuan, Kedis, Subuk, and Ularan Villages; very heavy soil erosion hazards (>480 t/ha/year) have been indicated at the Pujungan and Subuk Villages (Dewi et al. 2012). Land and water resources degradation seemly occurred because of land cover changes. Irrigated land use has significantly changed in the last two decades. Based on survey data at 20 Subak systems within the basin in 2014, approximately 37 % of paddy fields have decreased from 1223.42 ha in 1991 to 771.13 ha in 2014, converted to clove plantation (165.91 ha), vineyards (129.72 ha), dragon fruit farms (1.5 ha), fishponds (20 ha), buildings (housing, tourism facilities, offices, and trading facilities) (104.16 ha), and flooding areas of the Titab Dam (31 ha) (Budiasa et al. 2015). Land cover changes in the upstream, the Titab Dam construction at the center, and mining activities in the downstream brought heavy sedimentation into the river and irrigation channels, especially to the downstream area of the basin. Daily dumping of garbage and sewage into the river caused water quality degradation, indicated by saline surface water downstream. In fact, garbage in many points of the river and irrigation channels has actually blocked the water flows. These conditions changed fresh water sources for drinking water from the river to an Exclusive Water Supply Enterprise Perusahaan Daerah Air Minum (PDAM) or well water. The degradation also led to the scarcity of water supply for the new paddy fields under the Saba Irrigation Command Area, especially in the dry season.

The sustainable integrated water resources management goal at the Saba River Basin level is to improve the quality of life of a community based on implementation of local philosophy, called the Tri Hita Karana (three happiness causes). With regard to the GWP (2000), the IWRM involves the integration between natural system management and human system interaction. The natural system management involves the integration between freshwater and the coastal zone, land and water, surface and ground water, water quantity and water quality, as well as the integration of upstream and downstream areas of the basin. Furthermore, human system interaction embraces integration of all stakeholders in the planning and decision-making process, integrating water and wastewater management, and cross-sector integration (water for people, food, nature, industries, and other uses).

A physical effort to solve the development issues under the implementation of the IWRM concept was undertaken by the Titab Dam construction (2011–2015), supported by the government of Indonesia and funding of IDR 481,893,341,000. The project is expected to involve the following improvements: supply irrigation to the Saba and Puluran Irrigation Command Areas of 1794.82 ha; improve cropping intensity from 169 to 275 %; supply domestic water of 350 l/s in the Busungbiu, Seririt, Banjar, and Gerokgak Districts; supply electricity reserves of 1.5 MW in the Busungbiu District; and develop tourism, fishery, and conservation (BWS Bali-Penida 2011).

Moreover, it is very important to promote multistakeholders' participation in achieving implementation of the IWRM concept, including the optimization of the Titab Dam operation based on the *Tri Hita Karana* philosophy. A participatory approach has been implemented under an action research project entitled "Designing Local Frameworks for Integrated Water Resource Management," a collaboration between the Research Institute for Humanity and Nature (RIHN), Japan and Bogor Agriculture University (IPB), Indonesia (2012–2016).

5.2 Participatory Approach: A Principle in IWRM

The Global Water Partnership (GWP 2000) has promoted the four Dublin principles in IWRM involving the following: (1) fresh water is a finite and vulnerable resource, essential to sustain life, development, and the environment; (2) water development and management should be based on a participatory approach, involving users, planners, and policymakers at all levels; (3) women play a central part in the provision, management, and safeguarding of water; and (4) water has an economic value in all its competing uses and should be recognized as an economic good.

Based on the second principle, the participatory approach, water is a subject in which everyone is a stakeholder. Real participation only takes place when stakeholders are part of the decision-making process. Participation requires that stakeholders at all levels of the social structure have an impact on decisions at different levels of water management. A participatory approach is the only means for achieving a long-lasting consensus and common agreement. Participation is about taking responsibility, recognizing the effect of sectoral actions on other water users and aquatic ecosystems, and accepting the need for change to improve the efficiency of water use and allow the sustainable development of the resource (GWP 2000).

Since the high failure rate of the "top-down" or "blueprint" approach by governments, non-governmental organizations (NGOs), and international development agencies who designed programs without consulting intended beneficiaries, a participatory approach by participatory rural appraisal (PRA) is a fundamental ingredient in project planning (CIDE-WRI 1989). Since 1992, through the UN International Conference on Water and the Environment, a participatory approach has been promoted as an instrument that can be used to pursue an appropriate balance between a top-down and a bottom-up approach to IWRM. Governments at national, regional, and local levels have the responsibility for making participation possible, involving the creation of consultative mechanisms and participatory capacity, particularly among women and other marginalized social groups (GWP 2000).

A best practice in Indonesia for implementing of participatory approach is ensuring the multistakeholders' role in integrated water management and implementing an environmental services payment concept at the Cidanau watershed level, West Java. Based on this approach, a Communication Forum of the Cidanau Watershed was formally established on May 24, 2002, by the governor of Banten, decision number: 124.3/Kep.64-Huk/2002 (Rahadian 2013).

5.3 Multistakeholders' Meetings and Achievements

Multistages of a focus group discussion (FGD) method, under the participatory approach, implemented through several multistakeholders' meetings and committee meetings increased participation in IWRM. These FGDs focused on the key development issues and stakeholder identification, formulated a consensus and common agreements to formalize Community establishment, and established the role, function, and organizational structure/vision of the Community. Two stages of multistakeholders' meetings were held. The first multistakeholders' meeting was scheduled on September 8, 2013, at the Gran Surya Hotel in Seririt, Northern Bali. Ninety-nine participants from local governments, scientists, and local communities, such as the Subak system, attended this meeting. The purpose of the meeting was to introduce the research findings from 2012 to 2013, identify the general development issues, and establish the appropriate contributions of the stakeholders to IWRM within the Saba River Basin. The issues are gap rivers flow between the wet season and dry season, misuse and misdistribution of water, damaged irrigation infrastructure (weir and channels), existing buildings on the river border or banks of irrigation channels, leaks inside natural irrigation canals, increases in domestic water demand, land use changes from paddy fields for non-agricultural uses, low prices of rice at the farm gate, lack of integration, lack of a compensating mechanism between upstream and downstream, lack of coordination among the Subak systems that result in an increasing conflicts in water distribution, pest attacks, plant disease for rice farming, deforestation, sedimentation, garbage blockage inside the irrigation canals, mining activity within the hill area results in a rise of sedimentation in the wet season, and decreasing interest of youth in rice culture. In this condition, each stakeholder has partially responded to the appropriate issues. Furthermore, the second multistakeholders' meeting, which invited 132 stakeholders from local governments (at provincial and regency levels), scientists, business sectors, customary villages, Subak systems, and non-government organizations (NGOs), was held on October 24, 2014, at the same venue. Five resource speakers were invited from the following locations to improve local stakeholders' capacity to implement IWRM: (1) the Central Meteorology Climatology and Geophysics Agency, Jakarta; (2) the Indonesia Water Partnership, Jakarta; (3) the Forestry Agency of Bali Province, Denpasar; (4) the Communication Forum of Cidanau Watershed, Banten Province, West Java; and (5) the PT Krakatau Tirta Industry, West Java.

The achievements of the second multistakeholders' meeting include strength, weakness, opportunity, and threat identification (Table 5.1); formulation of key development issues; and a consensus for a Community establishment, as well the formulation of the Community Establishment Committee (CEC). Three stages of CEC meetings were held on December 11, 2014, at the Regional Development Planning Agency Office, Buleleng Regency, and on February 20, 2015, as well as on March 24, 2015, at the Seririt District Office. A tentative executive board meeting and a final multistakeholders' meeting were scheduled on September 23, 2015, and October 22, 2015, respectively. The first CEC (kickoff) meeting resulted in some decisions, such as a draft vision of the Community for IWRM, a draft of duties and functions of the Community, and stakeholder identification and mapping. The second CEC meeting decided the vision formulation as follows: "Realizing the sustainable Saba River Basin and the society welfare based on *Tri Hita Karana* philosophy," a type of Saba River Community, called as a forum, and the formulation of main duties and function of the Community. The final CEC meeting decided the priority of key development issues and draft action plans for garbage control and conservation; an organization structure of the Committee, including task forces;

Table 5.1 SWOT identification within Saba River Basin

SWOT Component	Details
Strength	Existing traditional institution: Subak system (including Subak Abian) and customary village
	Tri Hita Karana (three happiness causes) philosophy
	Availability of land and water resources for farming systems with many beautiful natural panoramas (river, natural pool, rice terrace, mountain, waterfall, sunrise, and sunset views)
	Potential production: Bali Coffee and Red Rice with local brand
Weakness	Daily behavior of local people to dump garbage into the river and irrigation channels
	Deforestation leads to the declining of water resources, increasing soil erosion hazard, and sedimentation
	Declining interest of rural youth to work as farmers, particularly as rice farmers because of too low income generation
	Traditional rules (*awig-awig*) of the Subak system are not powerful enough to prevent irrigated land use changes
Opportunity	Policy and regulation to implement detailed zoning, especially in the Seririt District
	Multipurposes of the Titab Dam
	Many foreign tourists come to beautiful natural panoramas within the Saba River Basin, potential for community-based tourism
	Local government policy for financial support in operation and maintenance of the Subak systems and customary villages
	Increasing interest of external demand to local products with good brands (Organic Bali Coffee and Organic Local Red Rice)
Threat	Climate changes result in flooding and drought
	Foreign or private investment encourages increasing land prices and irrigated land conversion for villas, hotels, housing, and trading facilities
	Mining activity in the downstream of the Saba River Basin leads to increasing soil erosion hazards and sedimentation

draft formulation of the basic rule Anggaran Dasar/Anggaran Rumah Tangga (AD/ART); and thirty tentative Executive Board members including nine members who serve as a management board of the forum.

The executive board member meeting held on September 23, 2015, at the Seririt District Office elected the management board and the additional executive board members with a total of 52 persons. This meeting decided the top two action plans (garbage control and soil and water conservation) and agreed with the basic rules (AD/ART) of the forum. Finally, an assembly meeting of the Saba River Basin Community was scheduled on October 22, 2015, at the Gran Surya Hotel, Seririt. The attendees agreed with the management board and executive board structure and selected the Community logo, temporary secretariat, and the top two priority action plans. At the final session of the meeting the environment board officer of Bali Province, as a representative of the governor of Bali Province, formally launched the Saba River Basin Community. This meeting was broadcast by Bali TV and published in a daily local newspaper of Bali Post.

5.4 The Saba River Basin Community Establishment

5.4.1 Vision, Main Duties, and Functions of the Community

The vision of the community initiated a kickoff meeting (first CEC meeting) at the Regional Development Planning Board Office of Buleleng Regency on December 11, 2014, and it was formulated during the second CEC meeting at the Seririt District Office on February 20, 2015. The vision formulation is "Realizing the sustainable Saba River Basin and the society welfare based on *Tri Hita Karana* philosophy." The *Tri Hita Karana*, the operational foundation of Bali's economic development, includes the following: (1) the harmonious relationship between human beings and God, as the creator of the universe; (2) the harmonious relationship among human beings themselves; and (3) the harmonious relationship between human beings and the environment (Regional Development Planning Board of Bali Province 2012).

The kickoff meeting also identified the main duties and functions of the Saba River Basin Community and mapped the existing stakeholders related to the basin. All stakeholders could be classified into four groups involving local governments, local communities, business sectors, and scientists. The main duties and functions of the Community were formulated at the second CEC meeting as follows: (a) to study the policy, planning, implementation, and outcomes of the Saba River Basin management, and the results of the study will be recommended to appropriate policy makers at provincial and regency levels; (b) to formulate annual action plans and report to appropriate policy makers at provincial and regency levels; and (c) to implement the following five roles and functions: as communication arena and exchange information across sectors and between governmental and other stakeholders within the Saba River Basin; to facilitate the coordination and negotiation process among stakeholders and/or sectors in formulating and implementing action plans; to stimulate problem solving by implementing action plans without overlapping or replacing the responsible technical agencies; to explore and empower local advantage by promoting local products with an interesting local brand, the uniqueness of culture heritage and creativity, and potential tourism within the Saba River Basin; and to build the local capacity through education and training, so they can improve their contribution to IWRM.

5.4.2 The Key Development Issues, Priority, and Action Planning

The general development issues were identified at the first multistakeholders' meeting, whereas key development issues were identified at the second multistakeholders' meeting. Furthermore, mapping and prioritizing of the key development issues was carried out at the third CEC meeting at the Seririt District Office on March 24, 2015.

Table 5.2 Key development issues, location, and priority

Key development issue	Location within the Saba River Basin	Priority and program 2015–2020
1. Dumping garbage and garbage blockage in the river and irrigation canals	Many points at upstream, central, downstream	First priority by program of garbage and environmental impact control
2. Deforestation and soil erosion hazard	Upstream (the Pujungan, Pupuan, Kedis Villages), central (the Subuk and Ularan Villages)	Second priority by program of natural resources conservation and environmental services
3. Heavy sedimentation in irrigation canals and damage of irrigation infrastructure	Downstream (Saba Irrigation Command Area)	Third priority by program of land use changes control and irrigation asset security
4. Irrigated land use changes for non-agricultural uses	Upstream, midstream, and downstream	Fourth priority by program of land use changes control and irrigation asset security
5. Many points of irrigation canals bank [border] are claimed as private assets	Downstream (the Saba and Puluran Irrigation Command Area)	Fifth priority by program of land use changes control and irrigation asset security
6. Declining interest of rural youth to work as farmers, particularly as rice farmers	Upstream, midstream, and downstream	Sixth priority by program of local advantage development and promotion
7. Agrotourism and ecotourism potential are not optimally developed yet	Upstream (Subak Munduk and Subak Sanda), central (Subak Titab, Subak Asah Uma, Titab Dam), and downstream (Subak Ponjokcukli)	Seventh priority by (1) program of local advantage development and promotion and (2) program of training, partnership, and collaboration
8. Local product branding not optimally developed yet	Upstream (Local Red Rice, Bali Coffee) and downstream (Haten Wine)	Eighth priority by (1) program of local advantage development and promotion; (2) program of training, partnership, and collaboration
9. Damage at the Saba River border	Downstream (Petemon, Seririt, Lokapaksa Villages)	Ninth priority by program of land use changes control and irrigation asset security
10. Increasing conflict in the use of water resources, for example, domestic water uses are directly from water spring	Upstream, central, and downstream of the Saba River Basin	Tenth priority by program of natural resources conservation and environmental services

Table 5.2 provides the location and priority of the top 10 key development issues within the Saba River Basin.

In 2015, based on Table 5.2, the Community focused on the top two issues, specifically garbage dumping and garbage blockage in the river and irrigation canals, as the first priority, and deforestation and soil erosion hazards, as the second priority. With regard to the first priority, the Task Force of the Garbage and Environmental Impact Control Program formulated action planning; with regard to the second, the Task Force of Natural Resources Conservation and the Environmental Services Program formulated action planning. The task forces were established at the third CEC meeting.

5.4.3 Membership, Organization Structure, and Funding

The membership, organizational structure, and Community funding were formulated at the third CEC meeting at the Seririt District Office, on March 24, 2015. The established Community of the Saba River Basin has the following structure: (1) protector, (2) advisory board, (3) management board, and (4) members.

Based on the basic rules of the Saba River Basin Community, a member is ex officio or representative of the stakeholder group if he hails from (1) the local government at provincial and regency levels; (2) the local community, such as the Subak (wetland Subak), Subak Abian (dry-land Subak), Subak Gede (Subak federation inside one irrigation command area), Majelis Alit of Subak (Subak federation inside one district), customary village, Majelis Alit of customary village (customary village federation inside one district), and non-government organizations (NGOs); (3) appropriate business sectors, such as an Exclusive Water Supply Enterprise (PDAM), National Electricity Corporation (PT PLN Persero), a bottled water company, Exclusive Market Enterprise (PD Pasar), a hotel association, and agribusiness association; and (4) scientists from an appropriate university and research institute. A member of the Saba River Basin Community must obey the basic rules, agree, and implement the basic principle and operational foundation as well as become aware and support the Community's activities. The membership will terminate in one of the following circumstances: (1) there is an urgent request by his origin group or institution, (2) death of the member who was not replaced by other competent members, and (3) the member was fired.

A protector provides protection for each effort by the Community in the decision-making process and implementing the IWRM concept, based on the basic principle "one river basin, one plan, one management" and the *Tri Hita Karana* philosophy toward a sustainable Saba River Basin and society welfare. The protectors include the governor of Bali Province, the Head of Buleleng Regency, and the Head of Tabanan Regency. The management board structure of the Community consists of the following: (1) the Head, (2) the Vice Head, (3) the Secretary, (4) the Treasurer, (5) the Executive Board, and (6) the Task Force Coordinators, as mentioned in Table 5.3. The Head of the Saba River Basin Community is elected by consensus or voting through an executive board meeting; however, he and the other members of the management board have to be initially elected by consensus through the CEC meeting. The official period of the management board is five years; however, this period can be extended through an agreement at an executive board meeting. A management board member will complete his job at the end of the official period although he finished his position early at the origin institution or stakeholder group. In managing the Saba River Basin, a Head is always controlled, advised, and evaluated

Table 5.3 Management board of the Saba River Basin Community

Position	Stakeholder
Head:	I Putu Nesa, SH (Head of the Customary Village Federation of Seririt District)
Vice Head:	I Gusti Putu Redana (Head of the Customary Village Federation of Gerokgak District)
Secretary:	I Putu Darsana (Secretary of the Customary Village Federation of Banjar District)
Treasurer:	I Gusti Bagus Putra Yasa (Head of the Subak Karangsari, Patemon Village, Seririt District)
Executive Board:	1. Head of the Regional Development Planning Agency of Bali Province
	2. Head of the Forestry Office of Bali Province
	3. Head of the Environmental Board of Bali Province
	4. Head of the Unda-Anyar Watershed Management Agency
	5. Head of the Bali-Penida River Basin Agency
	6. Head of the Meteorology Climatology and Geophysics Agency of Bali Province
	7. Head of the Public Works Office of Bali Province
	8. Head of the Food Agriculture and Horticulture Office of Bali Province
	9. Head of the Regional Development Planning Agency of Buleleng Regency
	10. Head of the Regional Development Planning Agency of Tabanan Regency
	11. Head of the Environmental Board of Buleleng Regency
	12. Head of the Environmental Board of Tabanan Regency
	13. Head of the Forestry and Plantation Office of Buleleng Regency
	14. Head of the Forestry and Plantation Office of Tabanan Regency
	15. Head of the Agriculture and Animal Husbandry Office of Buleleng Regency
	16. Head of the Agriculture and Horticulture Office of Tabanan Regency
	17. KASAT BINMAS Police of Buleleng Resort
	18. Head of the Police Pupuan Sector
	19. Head of the Seririt District
	20. Head of the Busungbiu District
	21. Head of the Banjar District
	22. Head of the Gerokgak District
	23. Head of the Pupuan District
	24. Coordinator of Perbekel (Village Head) Communication Forum, Pupuan District

(continued)

Table 5.3 (continued)

Position	Stakeholder
	25. Coordinator of Perbekel (Village Head) Communication Forum, Busungbiu District
	26. Coordinator of Perbekel (Village Head) Communication
	Forum, Banjar District
	27. Coordinator of Perbekel (Village Head) Communication Forum, Seririt District
	28. Coordinator of Perbekel (Village Head) Communication Forum, Gerokgak District
	29. Head of Customary Villages Federation of Pupuan District
	30. Head of Subak/Subak Abian Federation of Pupuan District
	31. Head of Subak Federation of Gerokgak District
	32. Coordinator of Subaks at Munduk Village
	33. Head of Subak Kekeran
	34. Head of Subak Titab
	35. Head of Subak Belumbang
	36. Head of Subak Pangkung Kunyit
	37. Head of Subak Abian Batur Pendem (Upstream)
	38. Head of Subak Abian Wiratani III (Upstream)
	39. Head of Subak Abian Wana Amerta (Upstream)
	40. Head of Subak Abian Ularan Village (Central)
	41. Head of Subak Gede Amerta Sari inside Busungbiu-Tunju Irrigation Command Area (Central)
	42. Main Director of PDAM, Buleleng Regency
	43. Director National Electricity Corporation, North Bali Area
Task Force Coordinator	
1. Garbage and Environmental Impact Control Program:	Ir. Gede Pariawan (Head of Sulanyah Customary Village)
2. Natural Resources Conservation and Environmental Services:	Gede Degdeg (Head of Subak Tinggarsari)
3. Local Product Development and Promotion:	Putu Witaya (Head of Customary Villages Federation of Busungbiu District)
4. Land Use Change Control and Irrigation Asset Security:	Nyoman Sudarma, S. Kep. (Head of Subak Federation of Seririt District)
5. Training, Partnership, and Collaboration Program Development:	Dr. I Wayan Budiasa, SP, MP (Faculty of Agriculture, Udayana University)

by an advisory board. Its members come from external parties involving the government, scientists, and the business sector. It has special authority when the management board has no power to control the existence of the community.

A management board meeting will be held at least three times, and executive board meetings and task force meetings will be held at least two times, annually. An assembly meeting, which invites all members, will be held at least two times, firstly, at the starting time and, finally, at the end of official period of the Community.

The Executive Board, as mentioned in Table 5.3, consists in 52 representative stakeholder groups (involving the Head, the Vice Head, the Secretary, the Treasurer, and the Task Force Coordinators) of the Saba River Basin Community. A Secretary and a number of members from Executive Board and/or other community members and/or competent invited third parties will assist the Task Force Coordinator. A number of active Task Force Coordinators will be flexible annually, depending on the appropriate action plans. For example, the only active coordinators in 2015 were the Task Force Coordinator for Garbage and Environmental Impact Control Program and the Task Force Coordinator of Natural Resources Conservation and Environmental Services Program. Each task force is responsible for formulating appropriate action plans that refer to key development issues.

Funding for the Saba River Basin Community operation can be provided from the national budget, the regional budget, and donors (other independent funding). Funds received and allocated refer to appropriate regulation. Financial support for establishing the Saba River Basin Community was received by the Research Institute for Humanity and Nature (RIHN), Japan, and Bogor Agricultural University (IPB), Republic of Indonesia, under the Research Project on Designing Local Frameworks for IWRM. Besides financial support, the Committee will be helpful, useful, and powerful when it obtains legal support from the governor of Bali Province or another appropriate agency.

5.5 Summary

A research project initiated by the Research Institute for Humanity and Nature, Japan and Bogor Agricultural University, Indonesia (2012–2016), was successful in encouraging and enhancing multistakeholders' participation in the Saba River Basin. In regard to IWRM, local communities have also improved their capabilities after multistages of multistakeholders' meetings, dissemination of research, best lessons learned by the Head Communication Forum of the Cidanau Watershed, and other resource speakers. SWOT conditions and key development issues in the Saba River Basin have been clearly identified by the meetings. The most important consensus achieved by the participatory approach was the establishment of the Saba River Basin Community based on "one river, one plan, one management" principle. Multistakeholders' participation achieved important common agreements, such as the following: (1) the vision of the Saba River Basin Community "Realizing the sustainable Saba River Basin, the society welfare based on *Tri Hita Karana* (three happiness causes) philosophy"; (2) the five basic roles and functions of the Community (communication, coordination, implementation of action plan, promotion, and education activities); (3) the organizational structure and positioning; (4) the basic rules of the Community; and (5) the top ten priority programs 2015–2020 and the top two action plans 2015/2016, i.e., garbage control and conservation within the Saba River Basin. The established Community was launched on October 22, 2015, by the Assembly (the main multistakeholders) meeting and broadcast by Bali TV and published in the Bali Post.

However, stakeholders are now standing at the starting line as they are at the stage where they have succeeded in establishing the Community. With this organization they have independently established, they must, in addition to deciding on a plan, transition into a concrete activity. The next stage demands that stakeholders push for action through a process of trial and error that builds the future.

Finally, the Saba River Basin Community will be more effective when it obtains appropriate funding support. It is very important for the Community to make successful partnerships, such as with the Indonesia Water Partnership, the Global Water Partnership, and other appropriate donors in addition to fostering collaboration with Central and Local Governments. Furthermore, it will be helpful to obtain legal support from the governor of Bali Province or another appropriate agency, since the Saba River Basin includes both the Buleleng and Tabanan Regencies.

References

Alaerts G, Muigne GL (2003) Integrated water management at river basin level: an institutional development focus on river basin organizations. World Bank

BPDAS [Watershed Management Agency] Unda-Anyar (2008) Land use modeling for basin management unit of Saba-Daya watershed. Denpasar. Rep, 73p

Budiasa IW et al (2015) The role of the subak system and tourism on land use changes within the Saba Watershed, Northern Bali, Indonesia. J ISSAAS 21(2):31–47

BWS [River Basin Agency] Bali-Penida (2005) Irrigation command areas within Saba River Basin. Denpasar, 1p

BWS [River Basin Agency] Bali-Penida (2011) Titab dam construction in Buleleng regency, Bali province, 16p

CIDE-WRI [Center for International Development and Environment of the World Resource Institute] (1989) Participatory rural appraisal handbook: conducting PRAs in Kenya. World Resources Institute, Center for International Development and Environment, Washington, DC

Dewi IGASU et al (2012) Erosion estimation and plan of land and water conservation at Saba watershed. E-J Trop Agroecotechnol 1(1):12–23

GWP [Global Water Partnership] (2000) Integrated water resources management. Sweden, 67p

GWP [Global Water Partnership] and INBO [International Network of Basin Organizations] (2009) A hand book for integrated water resources management in basins. Sweden, 103p

Hassing J et al (2009) Integrated water resources management in action. UNESCO, 18p

Nakagiri T et al. (2013) Basic diagnose of the present situation of water quality environment in the Saba river basin. Poster presentation on 1st Bali Stakeholder meeting on 8 September, 2013, C-09-Init Indonesia Chapter Project Collaboration between Research Institute for Humanity and Nature (RIHN) and Bogor Agricultural University (2011–2016)

Rahadian Np (2013) Towards integrated management of Cidanau watershed: multistakeholders role in managing integrated watershed and implementing the concept of environmental services payment. BPDAS Citarum-Ciliwung, Serang, 167p

Regional Development Planning Agency of Bali Province (2012) Economic development plan in Bali province. Paper presented at seminar on nature, culture and economic in Bali Island 5–6 September 2012 in collaboration with Research Institute for Humanity and Nature (RIHN)-Japan, Udayana University, Bogor University of Agriculture, and Hasanuddin University. Bali-Indonesia

Chapter 6
Hydrogeochemical Assessment of the Contribution of Caldera Lakes and Paddy Irrigation to River Water Stability

Takao Nakagiri, Hisaaki Kato, Seiji Maruyama, and Satoko Hashimoto

Abstract On-site water sampling surveys were carried out from 2012 to 2014 to sample rainwater, river water, and paddy water in the Saba River Basin in Bali Island and three caldera lakes adjacent to this basin, in order to characterize the different isotopic ratios in this range of water samples. The results showed that the water isotopic ratios of river water could be expressed by the linear combination of the isotopic ratios of the constituent waters such as rainwater and paddy water. Furthermore, the results suggested that there was likely a stable inflow from the caldera lakes, which were located outside of the basin area, to the Saba River via its tributaries (the Panas River and the Ling River). Based on the water isotopic relationship, the return flow from the paddy fields was inferred to contribute considerably to the river flow to the extent beyond its area ratio to the total basin area, especially in dry seasons. We concluded that the caldera lakes and the paddy irrigation practices in the Saba River Basin contributed to the stabilization of the river water flow and that this contribution was especially strong in the dry seasons. The water isotopic properties should be regarded as "scientific information for society," which can provide strong support for the diagnosis of river basins and decision-making for water resources management in the future.

T. Nakagiri (✉)
Graduate School of Life and Environmental Sciences, Osaka Prefecture University,
1-1 Gakuen-cho, Naka-ku, Sakai, Osaka 599-8531, Japan
e-mail: nakagiri@envi.osakafu-u.ac.jp

H. Kato
Research Institute for Human and Nature, 457-4 Motoyama,
Kamigamo, Kita-ku, Kyoto 603-8047, Japan

S. Maruyama
Kyoto Fission-Track Co., Ltd., 44-4 Oomiyaminamitajiri-cho,
Kita-ku, Kyoto 603-8832, Japan

S. Hashimoto
Center for Computational Sciences, Tsukuba University,
1-1-1, Tennodai, Tsukuba, Ibaraki 305-8577, Japan

K. Nakagami et al. (eds.), *Sustainable Water Management*,
DOI 10.1007/978-981-10-1204-4_6

Keywords Bali Island • Water stable isotope • Caldera lake • Paddy irrigation • River water stability

6.1 Introduction

We strongly depend on river water as a water resource for many uses, including agricultural, industrial, and domestic. River water stability, therefore, is very important from the view of water security and can also contribute to conserve biodiversity.

Although all river water originates from rainwater, there are many routes that the rainwater follows to rivers after arriving at the ground surface. Some rainwater will reach rivers relatively quickly by flowing along the surface, while other rainwater will infiltrate the surface or be stored in the closed water area like a lake and may take days or even years before discharging into the river. This is the reason why there are many rivers where water flow can be observed for several days or, in some cases, for more than half the year in areas with dry season during periods of no or little rainfall.

One of the major interests in hydrology is to clarify the routes and the ratios of different types of water that constitute river water, and this is also very important for effective water resources management. However, this information is very difficult to obtain, and effective methods to identify them have not been established yet.

In this situation, a methodology to use the stable isotopic ratios of water (herein, "water isotopic ratios") can be expected to solve this problem. Water consists of hydrogen and oxygen and mostly contains ^{1}H with a mass of 1 and ^{16}O with a mass of 16. Other stable isotopes, such as ^{2}H (also denoted as "D"), ^{17}O, and ^{18}O, also exist in trace amounts in nature. The existence ratios of these stable isotopes vary depending on the abundance of evaporation exposure and are also conservative. These mean that if the river water consists of several types of water that have obviously different evaporation experiences, then the existence ratios of the stable isotopes in river water can be expressed by a linear combination of each individual ratio. Therefore, there is a possibility that the constituent ratio of each type of water can be estimated in an inversely operating manner. In this chapter, this possibility is investigated through a case study in the Saba River Basin, located in the northern part of Bali Island, Indonesia. The extent to which caldera lakes and paddy irrigation practices contribute to the river flow stability is also discussed.

6.2 Overview of the Study Area

6.2.1 Geography and Land Use

The study area is the Saba River Basin located in the northern seaside of Bali Island, Indonesia. Bali Island (8.06°S–8.85°S and 114.43°E–115.71°E) is in tropical climate zone just to the south of the equator. The basin area is 131 km^2. Figure 6.1a shows the location and geographical shape of the basin. The Saba River flows into the

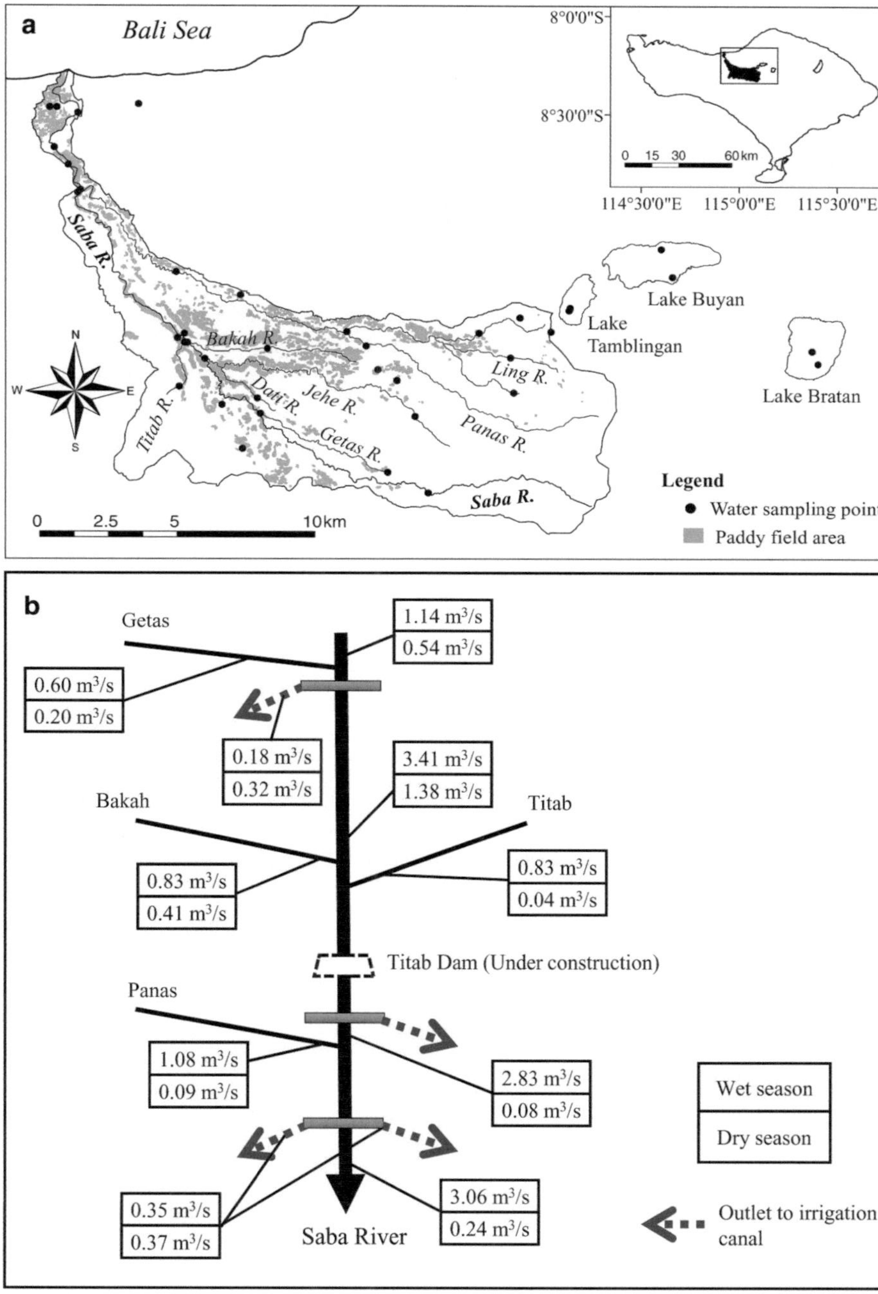

Fig. 6.1 Overview of the study area showing (**a**) a map of the Saba River Basin and location of the water sampling points and (**b**) the mean river flows in the wet and the dry seasons at each point of the Saba River. The values of river flow were the mean of the observed data in February 2013 and in February 2014 for the wet season and in August 2012 and in September 2013 for the dry season

northern sea from the mountainous area in the central part of the island and has many tributaries. Several representative tributaries are shown in Fig. 6.1a. There are three large caldera lakes (Bratan, Buyan, and Tamblingan) located to the east and just outside of the upper basin boundary. There are very few enclosed water areas such as irrigation ponds inside the basin.

The predominant land use in the basin is agriculture. In the lower area of the basin from midstream to downstream where the elevation is approximately <500 m above sea level, most of the farmland are paddy fields, and they are distributed over this area as shown in Fig. 6.1a. In the paddy fields in this basin, double or triple rice cropping through the year is commonly practiced. Although there is a dry season with little rainfall for several months in Bali Island, in many paddy fields, enough irrigation water is available mainly from the main or tributary rivers even during such periods.

In the highland area located upstream, where the elevation is >500 m above sea level, the farmland is mainly agroforestry, and tree crops such as coffee, cloves, and coconut are planted.

Intensive residential areas are limited in the downstream area.

6.2.2 Climatology and Hydrology

The study area belongs to a tropical climate zone and has two seasons: wet (from approximately November to June) and dry (July to October). According to the measurement records at the center part of the basin from 2013 to 2015 (Saptomo et al. 2015), the monthly rainfall was 200–380 mm/month in the wet season and 0–100 mm/month in the dry season. The average daily evapotranspiration rate is 3 mm/day. The annual fluctuation of solar radiation is small because Bali Island is located close to the equator, and even during the wet season, there are many days with enough solar radiation to evaporate water in the daytime. Consequently, the daily evapotranspiration rate is relatively stable throughout the year.

Several hydrological field surveys were conducted to obtain basic information about the water environment over the Saba River Basin. River flows at several joining points between main and tributary rivers were measured, and representative water samples were obtained to analyze the water quality. The parameters included pH, electrical conductivity (EC), suspended solids (SS), total nitrogen (TN), total phosphorous (TP), and chemical oxygen demand (COD).

Figure 6.1b shows the average river flows and intake flow to the main irrigation canal in the wet and dry seasons at each point of the Saba River.While in the dry season, the river flows were very stable, and in the wet season, they largely fluctuated due to the influence of rainfall. The river flows in the wet season shown in Fig. 6.1b were observed to be relatively stable without any heavy rainfall events just before the survey was taken. In the upstream and midstream areas, the river flows in the wet season were two to three times greater than in the dry season, while larger differences were observed in the downstream region. This was because the river water was diverted into several irrigation canals by an almost constant volume throughout the year.

There was no significant difference in the analytical results of the water quality throughout the year, except for the SS values. The values of pH and EC were 7.5–8.4 and 18–49 mS/m, respectively, in both the wet and dry seasons. The values of TN and TP were <1 mg/L and <0.05 mg/L, respectively, and the COD was 3–18 mg/L. In contrast, there was a significant difference in the values of SS between the wet and the dry seasons: where the SS was 8–63 mg/L in the wet season and <0.05 mg/L in the dry season. Even in the wet season with higher SS, the water quality of the Saba River was generally good, especially for irrigation use and to some extent for domestic use. Although the COD may have been lower than expected due to the large amounts of garbage dumped into the river in many places along the river stream, the river water flow only takes approximately 1 day from the uppermost stream to the river mouth; thus, the water quality likely does not degrade in the current situation.

6.3 General Description of the Water Stable Isotope

Isotopes are atoms which have the same properties of a chemical element but with different numbers of neutrons. Many atoms have several isotopes. Although some isotopes are very easy to decay to other isotopes spontaneously and are therefore "unstable," others do not apparently decay under natural conditions and are therefore "stable."

Hydrogen and oxygen, which are the elemental constituents of water, have two and three stable isotopes, respectively, in natural conditions: ^{1}H and deuterium (D or ^{2}H) and ^{16}O, ^{17}O, and ^{18}O. The mole fractions (existence ratios) of D, ^{17}O, and ^{18}O in nature are very small (e.g., $^{1}H{:}D = 0.99984426{:}0.00015574$; $^{16}O{:}^{18}O = 0.9976206{:}0.0020004$) (Coplen et al. 2002). Therefore, they are normally reported as δ values in units of parts per thousand (denoted as ‰ or per mil) relative to the standard. The δ value is calculated by

$$\delta = \left(\frac{R_{\text{sample}}}{R_{\text{standard}}} - 1 \right) \times 1000 \left[‰ \right] \tag{6.1}$$

where R_{sample} and R_{standard} are the ratios of the heavy isotope to the light one (e.g., $^{18}O/^{16}O$) in the sampled water and the standard, respectively. The Standard Mean Ocean Water (SMOW) distributed by the International Atomic Energy Agency (IAEA) had previously been used for the standard material; however, it is no longer distributed. Presently, the Vienna-SMOW (VSMOW) is distributed by the IAEA as the standard material of analyses of the water isotopic ratios of unknown water samples.

Craig (1961) reported that the relationship between $\delta^{18}O$ and δD in rainwaters sampled over the world can be expressed by the following equation:

$$\delta D = 8\delta^{18}O + 10 \tag{6.2}$$

This equation is called the "Global Meteoric Water Line" (GMWL, or just MWL). Although recent studies have reported that the slope and y-intercept values of the MWL are actually slightly different from 8 to 10 (e.g., Rozanski et al. 1993), in this chapter, these values were used as approximate values. The numerical value of the intercept of Equation 8.2 is called the deuterium excess (DE) (Dansgaard 1964). The DE value of the MWL is generally equal to 10, though the DE value derived from rainwater collected from a specific site is known to vary by region.

When rainwater experiences evaporation (kinetic fraction effect) on or near the surface, the relationship between $\delta^{18}O$ and δD will change, resulting in a line with a slope smaller than 8. Theoretically, fresh rainwater with almost no evaporation experience (where the equilibrium fraction effect is still dominant) and surface water with sufficient evaporation experience can be distinguished by knowing the properties of their isotopes or the slope of the line (Kendall and Caldwell 1998).

6.4 Sampling and Analytical Procedures

6.4.1 Sampling of Water

In this study, on-site surveys for water sampling were carried out in 6 periods: (1) from 1 to 2 of September 2012 (dry season), (2) from 28 of February to 1 of March 2013 (wet season), (3) from 10 to 11 of September 2013 (dry season), (4) from 13 to 17 of December 2013 (wet season), (5) from 20 to 22 of February 2014 (wet season), and (6) from 12 to 15 of August 2014 (dry season).

In the first three surveys during periods (1), (2), and (3), only the river water was sampled at 4 points on the Saba main stream and at 4 points on the 4 major tributaries (the Getas, the Bakah, the Titab, and the Panas), which are located just upstream from the confluence with the Saba main stream. In the last three surveys, additional water sampling points on the river streams (the Saba and the tributaries) were added. A total of 92 river water samples were obtained. The water samples were filtered with 0.2 μm-mesh membrane filters and reserved in glass vials (6 mL in volume).

In the surveys during periods (4), (5), and (6), paddy waters at the upmost, around the middle, and the lower end points in some paddy blocks, which consisted of several plots in different areas, were also sampled. A total of 26 river water samples were obtained.

Rainwaters were sampled in the surveys during periods (4) and (5), which were in the wet season. During these periods, plastic bottles were placed to collect rainwater at several points without any interception. The mouth of each plastic bottle was equipped with a plastic funnel through the use of a plastic ball to prevent the collected rainwater from evaporating. After rainfall, the bottles were collected within 1 day, and the collected waters were poured into glass vials in the same method as the river water samples. A total of 15 rainwater samples were obtained.

In the surveys during periods (4) and (6), water sampling was carried out on three caldera lakes: Lakes Bratan, Buyan, and Tamblingan. The water was sampled at the center point of the water surface area in each lake at 5 m depths from the water surface to the lake bed or to 50 m in depth. A total of 23 lake water samples were obtained. The lake water samples were also poured into glass vials in the same method as the river water samples.

The vials filled with the sample waters were cooled by coolant mediums in cooler boxes just after sampling. The vials were cooled in the refrigerator in the base camp close to the study site, and they were kept in the Styrofoam box in order to maintain low temperatures during transportation to Japan. The water samples were reserved in the sample storage kept at 5 °C in the Research Institute for Humanity and Nature (RIHN), and the analyses of the water isotopic ratios were carried out within 1 week after arrival.

6.4.2 Analyses of Water Stable Isotopic Ratios

Picarro L2120-i and L2130-i cavity ring-down spectrometers (CRDS) coupled with CTC LC-PAL liquid autosamplers were used for simultaneous analysis of the hydrogen and oxygen water isotopic ratios (i.e., δD and $\delta^{18}O$) of the water samples at the RIHN. In this study, the analyses were carried out according to the analytical procedures described by Maruyama and Tada (2014). Working standards distributed by Los Gatos Research, Inc. (hereafter abbreviated as LGR-WSs) were used for calibration of the analytical results. A set of three LGR-WSs (i.e., #3, #4, and #5), covering −79 to −9.8‰ of δD and −11.5 to −3.0‰ of $\delta^{18}O$ for calibration, were normally used for calibration of the measurement values of the water samples obtained from Bali Island. The standard errors ($1\sigma_{mean}$) of the analytical values of δD and $\delta^{18}O$ obtained in this study were typically less than ±0.2‰ and less than ±0.04‰, respectively. Additional details are described in Maruyama and Tada (2014).

6.5 Results and Discussion

The water isotopic ratios of the water samples from Bali Island are summarized in Fig. 6.2. The rainwater line, represented as “BRL” (Bali Rainwater Line) in Fig. 6.2, was based on data obtained by Kayane (1992). The BRL is defined as the following equation:

$$\delta D = 8\delta^{18}O + 17 \tag{6.3}$$

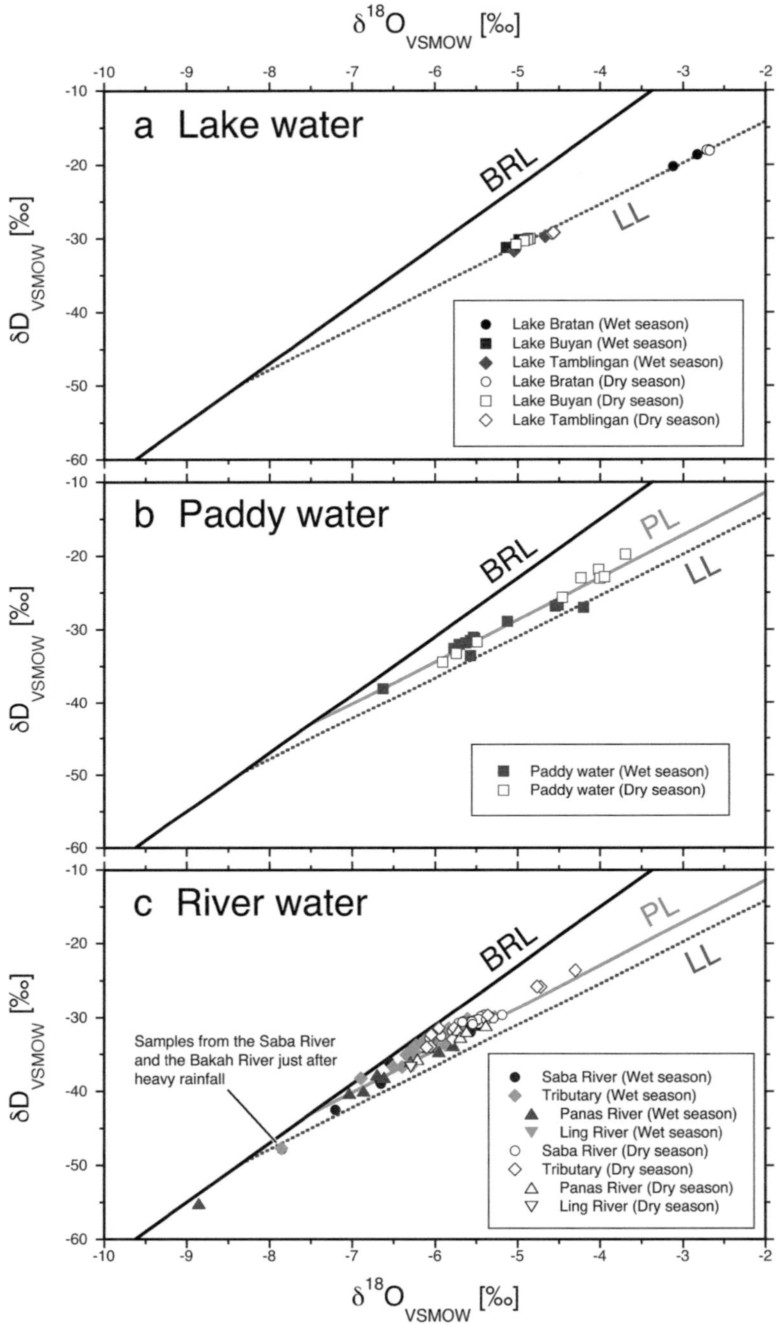

Fig. 6.2 The water isotopic ratios of (**a**) the caldera lakes, (**b**) the paddy waters, and (**c**) the waters obtained from the Saba River and its tributaries. The sizes of errors ($1\sigma_{mean}$) of the water isotopic ratios (δD and $\delta^{18}O$) of each sample were almost equivalent to or smaller than that of the symbol. Filled and open symbols represent the samples obtained in the wet and the dry seasons, respectively. The abbreviations are defined as follows: BRL is the Bali Rainwater Line, LL is the Lake Line, and PL is the Paddy Line. Though there were three plots in the region below the LL, two of the three overlapped around $\left(\delta^{18}O, \delta D\right) = \left(-7.9, -48\right)$

The water isotopic ratios of the rainwater samples obtained in the surveys for this study had a tendency to be slightly heavier than the BRL by up to ~1.5‰ in $\delta^{18}O$. In other words, the slope of the MWL obtained from the rainwater samples was slightly lower than that of the BRL, which had a slope of 8. The deviation may have been due to improper sampling of the rainwaters or alteration of the water isotopic ratios of the rainwater by evaporation during falling to the ground, as mentioned in the above section. The water isotopic ratios of rainwaters could potentially be altered by evaporation during falling to the ground (Dansgaard 1964; Hiyama et al. 2008). Therefore, the BRL should be employed as the MWL of Bali Island for the following discussions instead of the line obtained from the analytical values of the rainwaters sampled in this study.

Note that the water isotopic ratios of spring water in the upper area and waterfall water, which temporally appeared just after rainfall, were plotted just on the BRL.

6.5.1 Caldera Lake Water

In this study, the water samples were obtained from three of five caldera lakes in Bali Island: Lakes Bratan, Buyan, and Tamblingan. These caldera lakes have a closed system, and Shimano (1994) argued that there is no surface outflow from these lakes. The water areas of Lakes Bratan, Buyan, and Tamblingan are 383 ha, 471 ha, and 150 ha, respectively. The locational relationship of these lakes is shown in Fig. 6.1a.

The water isotopic ratios of the water samples obtained from the three caldera lakes are shown in Fig. 6.2a. The water isotopic ratios from each lake were similar to each other, and the $\delta^{18}O$ values ranged within 0.5‰. Moreover, the water isotopic ratios of each caldera lake obtained in the dry season were similar to those obtained in the wet season (Fig. 6.2a). All values of the water isotopic ratios of the caldera lakes are plotted along a single line, defined as the "Lake Line (LL)." The coefficient of determination (R^2) of the LL was almost 1.0, and the properties of the water isotopic ratios of the caldera lakes could be interpreted by using the LL.

The water isotopic ratios of Lake Buyan were similar to those of Lake Tamblingan, whereas those of Lake Bratan were ~2.0‰ heavier in $\delta^{18}O$ than Lakes Buyan and Tamblingan (Fig. 6.2a). Lakes Buyan and Tamblingan are isolated by cliffs that are >100 m high on the northern side. The cliffs can shade the water surface of these two lakes from the sun, and the moist air from the surfaces of the lakes is also difficult to exchange with relatively dry air on the surrounding land areas. On the other hand, moist air from the surface of Lake Bratan can relatively easily exchange with the surrounding dry air, and there is no shade on the surface of Lake Bratan in the daytime due to the absence of high cliffs surrounding the lake. Therefore, Lake Bratan can evaporate and experience the isotopic fractionation effect more strongly than Lakes Buyan and Tamblingan, resulting in water isotopic ratios that were heavier than the other two caldera lakes (Fig. 6.2a).

The water samples of Lake Bratan were obtained from the surface to 25 m in depth, those of Lake Buyan were obtained from the surface to 50 m in depth, and those of Lake Tamblingan were obtained from the surface to 20 m in depth. The water isotopic ratios of the water samples from various depths from each caldera lake had no statistical significant difference. Moreover, as shown in Fig. 6.2a, the water isotopic ratios of each lake obtained in the wet season had also almost no statistically significant difference from those obtained in the dry season. These results suggested that the water isotopic ratios of these three caldera lakes were maintained at a constant level throughout the year by advection and diffusion transport mechanisms in the lakes.

These caldera lakes may have experienced a dynamic evaporation effect for a long duration among the surface waters, which could have been incorporated into the river waters Therefore, the water isotopic ratios of each caldera lake were thought to be in an almost steady state, and the LL shown in Fig. 6.2a could be regarded as the convergence of the water isotopic ratios archived by the evaporation effect on Bali Island.

6.5.2 Paddy Water

The water isotopic ratios of the paddy waters obtained from the paddy fields in or near the Saba River Basin are plotted in Fig. 6.2b. The plots were observed to scatter more widely than those of the caldera lakes. Paddy rice cultivation requires a large amount of irrigation water throughout the cropping period, and a wide variety of waters, such as rainwater, groundwater, and return flow from paddy fields in the upper stream areas, could potentially be incorporated into the paddy water. Moreover, paddy water can be affected by the evaporation effect in varying degrees and can flow out to be incorporated into the river water or the groundwater again. Therefore, the paddy water and the water affected by paddy water may have highly variable water isotopic ratios. Considering this variability of paddy waters, the scattering of the water isotopic ratios of paddy waters seemed to be rather reasonable.

On the other hand, almost all of the plots in Fig. 6.2b were in the region between the BRL and the LL. The more the rainwater is affected by the dynamic evaporation effect, the heavier it becomes, and the plot determined by its isotopic ratios will diverge from the BRL and approach the LL; thus, the LL is regarded as a convergence archived by the evaporation effect in Bali Island. This result indicates that the water isotopic ratios of rainwaters affected by any amount of evaporation in natural conditions should be theoretically plotted in this region. Assuming that the paddy water was a general mixture of several waters, all of which originated from rainwater and had varying degrees of evaporation experience, and also received the dynamic isotopic fractionation effect by evaporation on the paddy plots, then almost all of the water isotopic ratios of the paddy waters would be expected to be plotted in this region.

The water isotopic ratios of the paddy waters obtained in both the dry and wet seasons were generally very close to the PL (Fig. 6.2b). This suggests that there may have been no remarkable difference in the isotopic fractionation process of the paddy waters between the dry and wet seasons.

While the water isotopic ratios of the paddy waters were scattered to some extent, they are still generally plotted along a single line, defined as the "Paddy Line (PL)" ($R^2 = 0.96$). We concluded that the properties of the water isotopic ratios of the paddy waters in the Saba River Basin could also be interpreted by using the PL, as well as the isotopic properties of the caldera lakes.

6.5.3 River Water

The water isotopic ratios of the samples obtained from the Saba River and its tributaries are plotted and distinguished by season in Fig. 6.2c. Note that for the discussion mentioned below, different styles of symbols were employed for the Panas River and the Ling River (one of the tributaries of the Panas River), though they are both tributaries of the Saba River (Fig. 6.1a).

In Fig. 6.2c, almost all of the plots are within the region between the BRL and the LL, and besides the majority of them are in the region between the BRL and the PL. Mathematically, if two kinds of waters with different water isotopic ratios, both of which are plotted in a region between two straight lines that are not parallel to each other, are mixed together, then the isotopic ratios of the mixed water would also be plotted within the same region. Considering the site situation that there was no significant surface water except for paddy water and surface discharge (like a temporal waterfall) just after rainfall inside the basin, most of river waters in this basin were assumed to basically consist of two kinds of waters that had the properties of the BRL or the PL.

On the other hand, in Fig. 6.2c, there are some plots in the region on or below the PL and above the LL. The water isotopic ratios above the PL could be explained by a linear combination of the meteoric waters and the paddy waters, whereas those on or below the PL could not be explained reasonably. If the water isotopic ratios of the river water were plotted just on the PL, then this means that all of the river water must have originated in the paddy water, though this is impossible in the actual situation of the Saba River. Moreover, if the ratios were below the PL, then no possible mathematical explanation exists. Almost all of the plots on or below the PL were from water samples obtained from the Panas River and the Ling River. The caldera lakes are located at the uppermost stream of the Ling River, though they are outside of the Saba River Basin boundary. Since these plots were still above the LL, they could be explained by incorporating the LL. For this reason, the water in the caldera lakes was assumed to have flowed into the Ling River and consequently into the Panas River beyond the basin boundary through the underground. Also, the quantity of the inflow was thought to be very stable because the river flows in the Ling and Panas Rivers in the dry season were very stable. Considering the terrain conditions, there appeared to be no inflow from the caldera lakes into the Saba River and the other tributaries, except for the Panas and Ling Rivers.

As shown in Fig. 6.1b, after joining between the Panas River and the main stream of the Saba River, the quantity of the Panas River flow accounted for 1/2 (in the dry season) to 1/3 (in the wet season) of the Saba River flow. The contribution of the Panas River to the Saba River flow was considerable, especially in the dry season. Consequently, the caldera lakes could possibly be contributing to the stabilization of the Saba River flow, especially in lower reaches in the dry season.

Three of the plots were out of the region between the BRL and the LL but very close to the BRL on the left side of the graph area in Fig. 6.2c. Note that two of the three plots were plotted around (δ^{18}O, δD) = (−7.9, −48) and thus were overlapped, so there appears to be only two plots in this region. Among these plots, one was obtained from the upper stream of the Panas, another was from the Bakah at the joining point between this tributary and the main stream of the Saba around the middle area of the basin, and the third was from the main stream of the Saba at the point very close to the second one. All of the water samples in these regions were taken while under the influence of large, direct runoff by consecutive heavy rainfalls from 13 to 16 December 2013 (93 mm in total, 43 mm/h at the peak). Therefore, rainwater accounted for a very large portion of these river water samples.

6.5.4 Relationship Between the River Water and Return Flow from Paddy Fields

As mentioned above, there appeared to be no area into which the caldera lake water flowed in the Saba River Basin, with the exception of the Panas River (including the Ling River) Basin, and the river water in the basin was assumed to be normally composed of the meteoric waters and return flow waters from the paddy fields. The meteoric waters contained not only "fresh" rainwater but also all of the different types of waters that had properties explained by the BRL such as the direct runoff (water that flows over the ground surface or through the ground directly into the river streams after rainfall) and the spring water.

Figure 6.3 shows the relationship between the ratios of the paddy field area to each sub-basin area (basin area of the tributaries), and the relative positions of the water isotopic ratios to the BRL and the PL by the season. We excluded the water isotopic ratios of the water samples from the Panas River and its tributary (i.e., the Ling River), which were assumed to be strongly influenced by the lake waters, and the water samples in which rainwater was considerably dominated by heavy rainfalls. The ratio r in Fig. 6.3 is defined as follows:

$$r = l_1 / \left(l_1 + l_2\right) \tag{6.4}$$

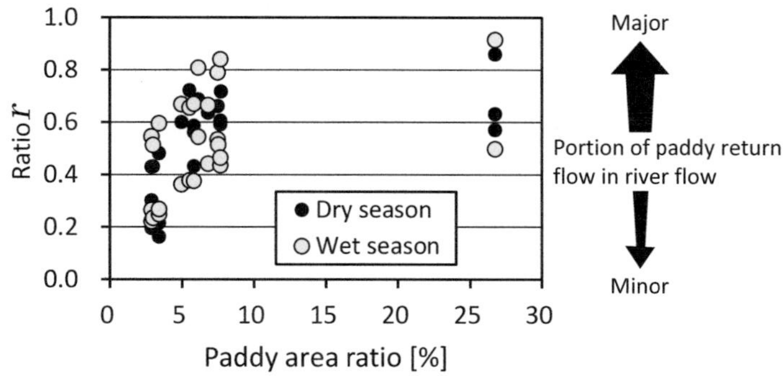

Fig. 6.3 The relationship between the area ratios of paddies in each sub-basin and the positions of the water isotopic ratios relative to the PL and the LL shown in Fig. 6.2

where l_1 is the distance from each plot to the BRL and l_2 is the distance from each plot to the PL in Fig. 6.2c. The larger the value of r for a certain plot is, the closer the plot approaches; namely, the constituent ratio of paddy water, which has the property of the PL, to the river water becomes theoretically higher.

In Fig. 6.3, for the paddy area ratios, the fluctuation range of the r values in the dry season when the river flow was relatively stable was smaller than that in the wet season. This suggested that the return flow from the paddy fields was stably contributing to the river flow.

On the other hand, for the small area ratios (approximately <5 %), the r values were also generally small (mostly $r < 0.5$), whereas for the larger area ratios (approximately <5 %), the r values were mostly >0.5, especially in the dry season (Fig. 6.3). Therefore, the paddy water was assumed to account for more than half of the river flow when the r value was >0.5. The paddy area ratios in each sub-basin in the Saba River Basin, however, were actually <27 %, indicating that the paddy fields above a certain area size could significantly contribute to the river flow by the return flow from them to extend beyond their area ratios.

6.6 Conclusion

In this study, the characteristics of the water isotopic ratios of rainwater, river water, and paddy water in the Saba River Basin in Bali Island and three caldera lakes adjacent to this basin were investigated. The water isotopic ratios of the samples obtained from Bali Island exhibited the characteristics according to the fundamental principle that all surface waters originated in the meteoric waters and that they are isotopically fractionated by evaporation effects (e.g., Alley and Cuffey 2001). The water isotopic ratios of rainwaters, caldera lake waters, and paddy waters could also be explained by specific linear isotope lines.

The relationship among these characteristics was investigated to reveal that the water isotopic ratios of river water could be expressed by a linear combination of the ratios. Furthermore, the results suggested that there was likely to be a stable inflow from the caldera lakes, which are located outside of the basin area, to the Saba River through its tributaries (i.e., the Panas and Ling Rivers). In particular, in the lower reach from the joining point between the Panas River and the main stream of the Saba, the contribution of the caldera lake water to the river flow appeared to be significant. Previous reports by Kayane (1992) and Shimano (1994) provided no evidence for a relationship between the caldera lakes and the neighboring rivers in Bali Island, despite their considerably detailed researches. As presented in this study, the water isotopic ratios of the natural water samples could be helpful for identifying the source and the routes of the river waters and evaluating the proportion of the individual water sources.

Except for the Panas River Basin and the area associated with it, the river water was likely composed of the meteoric water and the return flow from paddy fields. Based on the water isotopic relationship, the return flow from paddy fields considerably contributed to the river flow to the extent beyond its area ratio to the total basin area, especially in the dry season.

The topography of Bali Island is basically very steep and the lengths of all of the rivers on this Island are very short; therefore, the time required for water to flow on the ground surface from the topmost to the river mouth is only one or two days. In this situation, paddy rice agriculture allows rainwater and irrigation water from the river to remain in the basin for much longer times due to ponding in the paddy plots. This leads to an abundant return flow of water from the paddy fields even in the dry season. In other words, the water management in paddy fields, namely, paddy irrigation, can significantly contribute to the river flow.

In conclusion, the caldera lakes and the paddy irrigation in the Saba River Basin are contributing to stabilize the river water flow. Moreover, from the perspectives of not only natural science but also policy research where the results achieved by natural scientific research are applied, the isotopic properties of natural water samples can provide vital information for the conservation of river basins and related land resources. They provide a clear picture of the characteristics of water catchment areas through which waters from various sources such as rivers, paddies, and lakes pass. Moreover, they can contribute to demonstrating relationships between the control of sustainable development and water resources surrounding river basins, which may not have previously been considered. However, the development of social values from the isotopic analyses of natural water samples, in particular ideas for contributing to the formulation of policies, seems to have never been paid attention to until now. From this aspect, water isotopic properties should be regarded as scientific information for the improvement of society. We believe that they provide strong support for the diagnosis of river basins and decision-making for water resources management in the future.

References

Alley RB, Cuffey KM (2001) Oxygen- and hydrogen-isotopic ratios of water in precipitation: beyond paleothermometry. In: Valley JW, Cole DR (eds) Stable isotope geochemistry, Reviews in mineralogy and geochemistry 43. The Mineralogical Society of America, Washington, DC, pp 527–553

Coplen TB, Böhlke JK, de Bièvre P, Ding T, Holden NE, Hopple JA, Krouse HR, Lamberty A, Peiser HS, Révész K, Rieder SE, Rosman KJR, Roth E, Taylor PDP, Vocke RD Jr., Xiao YK (2002) Isotope-abundance variations of selected elements. Pure Appl Chem 74(10):1987–2017

Craig H (1961) Isotopic variations in meteoric waters. Science 133:1702–1703

Dansgaard W (1964) Stable isotopes in precipitation. Tellus 16:436–468

Hiyama T, Abe O, Kurita N, Fujita K, Ikeda K, Hashimoto S, Tsujimura M, Yamanaka T (2008) Review and perspective of the water cycle processes using stable isotope of water. J Jpn Soc Hydrol Water Resour 21(2):158–176 (in Japanese with English abstract)

Kayane I (ed) (1992) Water cycle and water use in Bali Island (Researchers related to the UNESCO's International Hydrological Programme in Japan). Institute of Geoscience, University of Tsukuba, Ibaraki, 320 p

Kendall C, Caldwell EA (1998) Fundamentals of isotope geochemistry. In: Kendall C, McDonnell JJ (eds) Isotope tracers in catchment hydrology. Elsevier, New York, pp 51–86

Maruyama S, Tada Y (2014) Comparison of water isotope analysis between cavity ring-down spectroscopy and isotope ratio mass spectrometry. Geochem J 48:105–109

Rozanski K et al (1993) Isotopic pattern in modern global precipitation. In: Swart PK et al (eds) Climate change in continental isotope records, vol 78, Geophysical monograph series. AGU, Washington, DC, pp 1–36

Saptomo SK, Chadirin Y, Setiawan BI, Wayan BI, Kato H, Kubota J (2015) Quantifying water balance of Subak paddy field based on continuous field monitoring. Jurnal Teknologi (Sci Eng) 76(15):53–59

Shimano Y (1994) Characteristics of hydro-geomorphology in the volcanic island of Bali, Indonesia. Bull Bunsei Univ Art 6:7–35 (in Japanese with English abstract)

Chapter 7
Reconsideration of the Meaning of Dam Construction for Water Resources Management: The Environmental Impact Assessment of the Titab Dam Project Toward Futurability of the Saba River Basin

I Wayan Budiasa, Hisaaki Kato, and Ken'ichi Nakagami

Abstract The authors consider the meaning of dam construction in the context of water resources management, based on the Titab Dam construction case study in Indonesia. The Titab Dam was built to supply irrigation water for 1794.82 ha of downstream irrigated paddy fields, provide domestic water for 72,263 potential household connections, reserve 1.5 MW of electricity, conserve water resources within the Saba River Basin, and develop fishery agro-eco-tourism activities. Besides the economic and environmental benefits, the construction of the Titab Dam also brought adverse impacts, including the disappearance of Subak Dukuh after the Subak Temple was destroyed; inundating of all paddy fields in Subak Dukuh (13 ha); flooding almost all of the paddy fields in Subak Asah Uma (18 ha); and destroying farmers' livelihood inside both Subak Dukuh and Subak Asah Uma. To achieve optimal multi-functionality and effectiveness of the Titab Dam, the top two priorities are implementation of garbage control and soil and water conservation action plans, as carried out by the Saba River Basin Community.

Keywords Titab Dam • Environment impact • Multi-functionality • Subak disappearance • Saba River • Water conservation

I.W. Budiasa (✉)
Faculty of Agriculture, Udayana University, Jalan Kampus Bukit Jimbaran, Jimbaran, Kuta Selatan, Kabupaten Badung, Bali 80361, Indonesia
e-mail: wba.agr@unud.ac.id

H. Kato
Research Institute for Humanity and Nature, 457-4 Motoyama, Kamigamo, Kita-ku, Kyoto 603-8047, Japan

K. Nakagami
College of Policy Science, Ritsumeikan University, 2-150 Iwakura-cho, Ibaraki, Osaka 567-8570, Japan

K. Nakagami et al. (eds.), *Sustainable Water Management*,
DOI 10.1007/978-981-10-1204-4_7

7.1 Introduction

The master plan for the Acceleration and Expansion of Indonesia's Economic Development 2011–2025 has named Bali-Nusa Tenggara as one of Economic Corridor, a part of its assignment as a "Gateway for Tourism and National Food Support" (www.indonesia-investments.com).

The economic growth of Bali depends on primary production (agriculture), secondary sectors (small-scale industries), and tertiary sectors (including tourism), totaling approximately 8.08 %, 15.57 %, and 66.35 % of the total growth, respectively (Regional Development Planning Agency of Bali Province 2012). The Bali Province, which consists of more than 563,666 ha, comprises 81,744 ha (14.5 %) of paddy fields that are traditionally managed through 1548 Subak systems, 273,655 ha (48.55 %) of other agricultural fields, and 208,267 ha (36.95 %) of non-agricultural land (http://bali.bps.go.id).

The Saba River Basin in Northern Bali covers an area of approximately 132.89 km^2 (BWS Bali-Penida 2011a). It is part of both the Buleleng and the Tabanan Regencies involving approximately 4156.82 ha of irrigated paddy fields, which is traditionally managed by 55 Subak systems and supported by more than eight sub-rivers (the Saba, the Getas, the Dati, the Jehe, the Bakah, the Titab, the Panas, and the Ling) for supplying irrigation water inside the area (Budiasa et al. 2015). It also covers part of five districts and 35 village areas. The population inside the 35 villages was 124,935 persons (http://bulelengkab.bps.go.id). Based on data provided by Balai Pengelolaan Daerah Aliran Sungai (BPDAS) Unda Anyar (2008), land formation within the Saba River Basin is mostly volcanic (upstream and central of the basin), and the coastal area is fluvial. The land slope is mostly 15–25%. The soil in the basin is physically classified as gray regosol, gray-brown andosol, and yellow-brown latosol. In the upstream, the soil is classified as gray-brown regosol. The soil in the central area is classified as gray-brown regosol, brown latosol, and latosol. Downstream, the soil consists of almost all gray-brown regosol and a small amount of gray-brown alluvial. Land cover in the basin is mostly land used for mixed farming systems, and the upstream land cover consists of primary forest, paddy fields, dry land farming, and settlements. Further, the land in the basin is categorized as non-critical land, potential critical land, and light critical land, in percentages of 17.55 %, 51.04 %, and 31.40 %, respectively (Dewi et al. 2012).

The research project, a collaboration between the Research Institute for Humanity and Nature (RIHN), Japan and Bogor Agricultural University (IPB), Indonesia, confirmed that the water quality within the Saba River Basin is acceptable for irrigation and daily domestic use, even downstream near the river mouth, as well as upstream. River flows at the water sampling points are shown in Fig. 7.1 (Nakagiri et al. 2013). Based on the figure, with the exception of the Panas River, river flows in the wet season were approximately four times that of the dry season (Budiasa et al. 2015). During the period of 1995–2004, the annual rainfall inside the basin tended to decrease with time. Before the great El Niño in 1997, rainfall reached 2500 mm

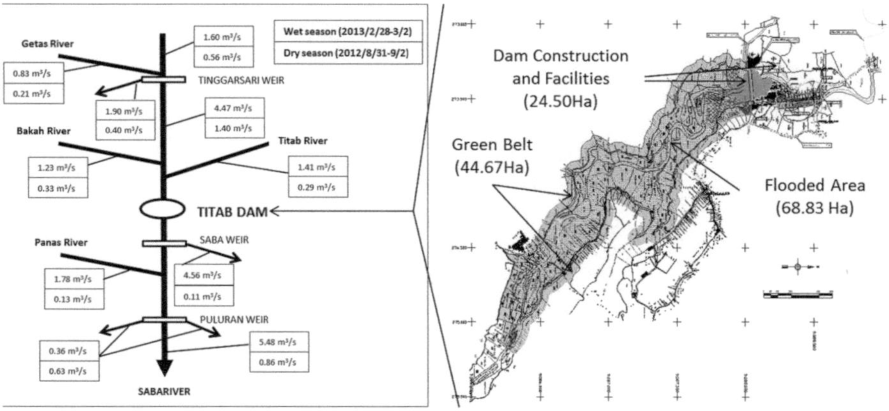

Fig. 7.1 River flow at the water sampling points (Nakagiri et al. 2013) (Left) and Titab dam location and site plan

in 1995; however, it decreased significantly and reached below 1500 mm (Setiawan et al. 2013).

The Saba River Basin has flooded several times. On December 31, 2010, an extremely severe flooding event occurred at the Titab Sub-river and the Saba River, when a heavy rain totaling approximately 48 mm (2037 l/s) was brought upon the Pucaksari and the Titab villages in only 3 h. It destroyed infrastructure, such as the Titab I, the Titab II, the Bukit Pulu, and the Batu Megaang weirs, the Titab bridge, the access road to the Batu Megaang sub-village, and 20 ha of agricultural land. The flood also claimed two lives and animals, such as Bali cattle (BWS Bali-Penida 2011a).

Some efforts to prevent flooding include the Saba River Basin rehabilitation, about 1905 m of primary irrigation channel and 903 m of secondary irrigation channel in 2012, and the implementation of the Titab Dam Project (2011–2015) by local and central governments (BWS Bali-Penida 2011a). The Titab Dam Project was initiated by BWS Bali-Penida, the Directorate General of Water Resources, the Ministry of Public Works with Implementation Constructor of PT Nindya Karya – PT Brantas Abipraya KSO (Contract No.: HK.02.03/PJSA-BP/WT/002 issued on April 12, 2011, from April 12, 2011 to December 31, 2014, with an investment cost of IDR 468,716,771,000), and consulting supervision by PT Indra Karya – PT Wahana Adya KSO (Contract No.: HK.02.03/PJSA-BP/WT/001 issued on April 1, 2011 for 1365 calendar days, from April 1, 2011 to December 29, 2014, with supporting funds of IDR 13,176,570,000). This largest of the six dams in Bali was finished and flooded on December 13, 2015. The impounding was formally implemented by Megawati Soekarno Putri (Ex-President of Indonesia), accompanied by the Ministry of Public Works and Settlement, the Head of Buleleng Regency, the Head of BWS Bali-Penida, and a number of other officers at local and central governmental levels.

This dam was built to regulate 1794.82 ha of irrigated areas of Saba and Puluran downstream of the basin in order to accomplish the following: to increase cropping intensity from 169 to 275 %, to ensure a domestic water supply of 350 l/s for the Seririt and Banjar district areas, to provide approximately 1.5 MW of electricity reserves for the Busungbiu district area, to promote tourism development, to conserve the Saba River Basin, and to develop fishery activities (BWS Bali-Penida 2011b).

This chapter is focused on the Environmental Impact Assessment (EIA) of the Titab Dam Project toward sustainable integrated water resource management (IWRM) in the Saba River Basin. Appropriate primary and secondary data have been collected during the research project entitled "Designing Local Frameworks for IWRM," as a collaboration between RIHN and IPB (2012–2016). Data analysis is based on a descriptive qualitative analysis method to address the EIA key issues.

7.2 Environment Impact Assessment Principles

The Environment Impact Assessment Principles is a tool for decision-makers to identify potential environmental impacts of proposed projects, evaluate alternative approaches, and design and incorporate appropriate prevention, mitigation, management, and monitoring measures. An EIA cannot be divorced from the social impact of the project; hence, the latter is considered as a key dimension of the EIA process. The EIA shall address both positive and negative potential environmental impacts of the given project, any related social implications, as well as eventual transboundary effects. An EIA evaluates a project's potential environmental and social risks and impacts in its area of influence. There are three environmental categories for field operations, labeled as A, B, and C (FAO 2011).

Category A projects involve significant, cumulative, or even potentially irreversible negative environmental impacts or risks. Typically, such projects include planned interventions that change existing land and/or water uses, open up new lands, disturb natural habitat needed for maintaining biodiversity, involve significant expansion of industry, introduce water impoundment schemes, promote the use of agrochemicals, or require the acquisition of land and/or resettlement of local populations. River basin development projects are indicated as category A projects. Category B projects do not entail significant (or potentially irreversible) negative environmental (and associated social) impacts, but may still have adverse effects, which can be mitigated with suitable preventive actions. Category B projects do not require a full EIA but will require further deepening of environmental or social considerations, depending on the expected magnitude of risks. Watershed management or rehabilitation, river basin management planning, international water management, and agreements for medium-size projects are indicated as category B projects. Furthermore, category C projects have minimal or no potential negative environmental (or social) impacts, either individually or cumulatively. They are not controversial in terms of the interests of key stakeholders. As such, they do not require further analysis or an impact assessment. Natural resource (including water resources) assessments and monitoring are indicated as a category C project (FAO 2011). Table 7.1 presents the environmental screening for category A or B projects.

Table 7.1 Environmental screening for category A or B projects

Would the project	Yes	No	Unable to determine
1. Does it have significant adverse impacts on public health or safety?		√	
2. Does it have significant or controversial environmental effects on biophysical resources, such as land, water, soil, biodiversity?		√	
3. Does it have adverse impacts on unique characteristics, such as wilderness, natural rivers, aquifers, prime farmlands, wetlands, floodplains, or ecologically significant areas?		√	
4. Does it have adverse impacts on traditional practices or agricultural systems in the area?	√		
5. Does it have highly uncertain and potentially significant environmental and social impacts with unique or unknown risks?			√
6. Does it establish a precedent for future action or represent a decision in principle about future actions with potentially significant environmental and social impacts?		√	
7. Does it set in motion or contribute to a progressive accumulation of significant environmental and social impacts?		√	
8. Does it have adverse impacts (direct or indirect) on natural habitats such as wetlands, mangroves, tropical forests?		√	
9. Does it have adverse impacts on important national or international species (listed or proposed) or on critical species habitats?		√	
10. Does it have adverse impacts on local or indigenous populations residing in the area of interest?		√	
11. Does it contribute to introduction, continued existence, or spread of non-native invasive species or promote the introduction, growth, or expansion of the range of non-native invasive species?	√		
12. Does it threaten national, local, tribal, or indigenous peoples' requirements for use of natural resources or protection of the environment?		√	
13. Does it trigger or exacerbate unresolved land tenure conflicts concerning rights or alternative uses of natural resources?	√		
14. Does it have a disproportionate, significant adverse effect on low-income or disadvantaged populations?		√	
15. Does it restrict access to traditional or ceremonial sites or adversely affect the physical integrity of such religious sacred sites?		√	
16. Does it have adverse impacts on natural resources or properties of historic or cultural significance?	√		
17. Does it lead to significant impacts indicated by a national, district, or local community group?		√	

(continued)

Table 7.1 (continued)

Would the project	Yes	No	Unable to determine
18. Does it have the potential to be controversial because of stakeholder disagreements?		√	
19. Does it encourage migration or other population shifts?		√	
20. Does it increase the workload of local communities or subgroups within the communities?	√		
21. Does it work in opposition with ongoing socio-economic development goals or efforts?		√	
22. Does it require capacity development of affected or involved individuals and organizations? Does it require capacity development to review and update of policies, laws, and regulations or to develop partnerships?	√		
Please answer the following questions	Yes	No	
1. Are the personnel preparing this form familiar with the site?	√		
2. Are the personnel familiar with the populations living in or near the site?	√		
3. List the name of those who have conducted or will conduct site visits and the dates: Dr. I Wayan Budiasa (2012–2016); Dr. Hisaaki Kato (2012–2016); Prof. Dr. Ken'ichi Nakagami (2012–2013)			
Certification of Project Category A or B	Yes	No	
I affirm the completion of an analysis of the potential environmental and social impacts for this project and certify it to be in category B. The analysis included information to assess the potential negative and positive impacts and is addressed in the project design through appropriate prevention or mitigation measures.	√		
I affirm the completion of an analysis of the potential environmental impacts and have determined this project should be classified as category A.		√	
Title, name, and signature of project leader:			
Dr. I Wayan Budiasa			
Date: December 31, 2015			

7.3 EIA Key Issues: Reconsideration of the Titab Dam Project

The implementation of the Titab Dam Project was scheduled from April 12, 2011 to December 31, 2014. However, the project was finished about 1 year later and the impounding of the dam occurred on December 13, 2015. This work began with land acquisition; infrastructure construction, such as access road and facilities; and dam construction. The project is located on land that is part of the Ularan, Ringdikit, Busungbiu, Telaga, and Titab villages, Seririt and Busungbiu districts, and Buleleng Regency. It required approximately 138 ha of land, including a flooded area of 68.83 ha, a green belt of 44.67 ha, and 24.5 ha allocated for infrastructure and facilities construction (Fig. 7.1). The time schedule for land acquisition is mentioned in Table 7.2. Based on the table, the total land acquisition was only 109.34 ha. The remaining land requirement of approximately 28.66 ha was contributed by a government asset as part of the Saba River body. Based on final negotiations between project management and representatives of the land owner, the average land price was IDR 100,000 for 1 m^2, including building compensation.

Based on the Rapid Rural Appraisal method, approximately 31 ha of paddy fields were contributed by Subak Dukuh totaling approximately 13 ha. Subak Asah Uma is about 18 ha and has been lost under the waters of Titab Dam (Budiasa et al. 2015). As depicted in Table 7.1, this project has adverse impacts on traditional practices or agricultural systems in the area. In the case of the land use change of the paddy fields because of the Titab Dam construction, some farmers who would like to continue their rice culture outside the project location are facing hardships because the land available for purchase is less than the one they sold. The disadvantage is a result of increasing land prices during/after project construction.

As depicted in Table 7.1, the Titab Dam Project has exacerbated unresolved land tenure conflicts concerning rights or alternative uses of natural resources. A number of farmers under the land tenure (profit-sharing) system face an opportunity cost because of the government intervention to convert the paddy fields and/or dry land farming to a flooded area. They may be looking for new livelihood with more effort, additional cost, and time expenditures learning a special skill. This condition might increase the workload of local communities or subgroups within the communities (see Table 7.1).

Based on Table 7.1, the entire extension dedicated to paddy fields inside Subak Dukuh were flooded following the project causing the Subak system to disappear altogether. The Subak Dukuh Temple was also destroyed that once held special Hindu ceremonies as an implementation of Tri Hita Karana philosophy at the Subak level. The organization structure, management board, and membership of the Subak have also dismissed itself. This means the project has adverse impacts on natural resources as well as properties of historic or cultural significance.

Furthermore, based on Table 7.1, this project has inspired multistakeholders' participation in implementing the IWRM concept within the Saba River Basin, by firstly establishing the Saba River Basin Community under a participatory approach (see Part 3–8). Their improvement was accomplished by multistakeholders meetings to achieve common agreements, and by dissemination of IWRM concepts and global best practices. Based on its vision entitled "Realizing the sustainable Saba River Basin, the society welfare based on *Tri Hita Karana* (three happiness causes) philosophy," it started to prioritize the top two action plans. Specifically, those action plans involve garbage control followed by soil and water conservation.

Based on the multi-functionality of the Titab Dam, it may serve as an important barometer toward the futurability of the Saba River Basin. To achieve the effectiveness of the Titab Dam operation, good land cover conditions upstream of the basin, best management practices in farming systems, and river free from garbage and wastewater are crucial. With regard to research by Dewi et al. (2012), 51.04% of

Table 7.2 Time schedule of land acquisition for the Titab Dam construction (upside), technical data of the Titab Dam construction (middle), and the estimated domestic water supply project under the Titab Dam operation (downside)

Time schedule of land acquisition for the Titab Dam construction								
MOU		2010 (Ha)		2011 (Ha)	2012 (Ha)	2013		Total land acquisition (Ha)
No. HK02.03-DA/558								
No. 075/15/KB/B.PEM/2010								
No. 075/479/PEM/2010								
July 6, 2010								
Gov. level	%	Land acquisition plan (Ha)				Area (Ha)	Cost (IDR)	
Central government (Ministry of Public Works)	50	69	–	9.50	–	32.43	32.78 billion	41.93
Bali Province	35	48.3	4.025	4.30	–	38.97	39.78 billion	47.295
Buleleng Regency	15	20.7	0.975	5.50	–	13.64	13.64 billion	20.115
Total	100	138	5.00	19.30	–	85.04	86.20 billion	109.34
Source: BWS Bali Penida (2011b)								

Technical data of Titab Dam		
Dam	Type	Random stone backfill with upright core
	High (m)	59.8
	Length (m)	210
Bin	Body volume (10^3 m^3)	12,800
	Effective volume (10^3 m^3)	10,080
	Flooded area (Ha)	68.83

Water discharge	Benefit and quantity	Irrigation: 1794.82 Ha	
		Domestic water supply: 350 l/s	
		Suppletion of interconnecting channel: 350 l/s	
Saba River Basin (Km^2)	132.89		
Spillway	Capacity (m^3/s)	385.09	
	Type	Side spillway without window	
Source: BWS Bali-Penida (2011a)			
The estimated domestic water supply project			
Networks area by district	Average of family size[a]	Potential number of household connection[b]	Estimated domestic water demand (m^3/year)[c]
Banjar	3.3	9099	1.20×10^6
Busungbiu	3.9	12,166	1.90×10^6
Gerokgak	3.7	30,665	4.55×10^6
Seririt	3.6	20,333	2.94×10^6
Total	3.65	72,263	10.59×10^6

[a]BPS Buleleng Regency (2013)
[b]BWS Bali-Penida (2012)
[c]JICA (2006)

land within the Saba River Basin was categorized as potential critical land; 31.40 % was categorized as light critical land. With regard to erosion, severe erosion was categorized in ranges from 192.02 to 403.63 t/ha/y in the Pujungan, Pupuan, Subuk, Ularan, and Kedis villages, totaling about 1852.34 ha. Extremely severe erosion that ranged from 545.97 to 728.60 t/ha/y was categorized in the Pujungan and Subuk villages, totaling about 1049.94 ha.

Planning of soil and water conservation is achieved by planting cover crops, storied canopies, and construction of terraces, starting with land combating extremely severe erosion and severe erosion. The conservation action plan is also prepared for the green belt of the Titab Dam, which is about 44.67 ha. The good land cover is expected to achieve the optimal water recharge, so there will be a stable water discharge between the wet season and dry season.

Water discharge from the Titab Dam is used to support 1794.82 ha rice culture downstream of the basin, consisting of 1396.4 ha and 398.42 ha, for the Saba and Puluran irrigation command areas, respectively (Table 7.2). It can be expected to improve cropping intensity from 169 % before the project to 275 % during the project operation, annually. It is also planned to ensure a domestic water supply of 350 l/s. The estimated domestic water demand under the Titab Dam Project is depicted in Table 7.2. With the assumption that per capita water consumption is 110 l/day or 40.15 m^3/y (JICA 2006), the average family size is 3.65 persons, and the total potential number of household connections is about 72,263, then we estimate the total domestic water demands will be about 10.59 million m^3/y.

The Gerokgak district has the largest potential number of household connections, approximately 30,665 units, with average of family size of 3.7 persons, requiring approximately 4.55 million m^3/y. It is followed by the Seririt, Busungbiu, and Banjar districts, with a potential number of household connections of 20,333, 12,166, and 9099 units and average of family size of 3.6, 3.9, and 3.3 persons, respectively. If, the domestic water supply system with a capacity of 350 l/s is fully operational, the total domestic water supply will reach to 11.04 million m^3/year. This is only 4.2% more than the total estimated water demand.

The other economic benefit of the Titab Dam Project is the reservation of 1.5 MW of electricity, to develop a fishery system and appropriate tourism destinations. This power reserve can support approximately 1666 household connections with an average power capacity of 900 W. The fishery system, which is developed inside the Titab Dam, is stocking fish periodically. An agro-eco-tourism package inside the Saba River Basin, involving the Titab Dam destination, will be interesting.

The Melanting Waterfall and the Tamblingan Lake upstream of the basin are presently the best destinations for tourists. Diving and other marine activities and sunset views downstream of the basin are also very noteworthy. Tracking, fishing, and water sports will be enticing attractions when offered to future tourists inside and surrounding the Titab Dam.

7.4 Summary

The sustainable Saba River Basin management and multi-functionality are the main considerations of the Titab Dam Project implementation. The EIA is a useful tool for conducting, monitoring, and evaluating the environment impacts and related social implications of the project's implementation. Besides the economic and environmental benefits, construction of the Titab Dam still has adverse impacts, such as the disappearance of the Subak Dukuh since the Subak Temple has been destroyed, all paddy fields in Subak Dukuh (13 ha) and almost all the paddy fields in Subak Asah Uma (18 ha) have been converted to flooded areas, and a number of farmers inside both the Subak Dukuh and Subak Asah Uma have lost their livelihood. To achieve the optimal multi-functionality and the effectiveness of the Titab Dam, the top two priorities are the implementation of garbage control and soil and water conservation action plans, as carried out by the Saba River Basin Community.

References

BPDAS [Watershed Management Agency] of Unda-Anyar (2008) Land use modeling for basin management unit of Saba-Daya watershed. Denpasar. Rep, 73p

Budiasa IW et al (2015) The role of the Subak system and tourism on land use changes within the Saba watershed, Northern Bali, Indonesia. J ISSAAS 21(2):31–47

BWS [River Basin Agency] of Bali-Penida (2011a) Technical data of Titab Dam construction. Denpasar, 1p

BWS [River Basin Agency] of Bali-Penida (2011b) Titab dam construction in Buleleng regency, Bali province. Denpasar, 16 p

BWS [River Basin Agency] of Bali-Penida (2012) Detail design of domestic water supply system development. Denpasar, 1p

Dewi IGASU et al (2012) Erosion prediction and planning of soil and water conservation at Saba Watershed. E-J Trop Agroecotechnol, Faculty of Agriculture Udayana University 1(1):12–23

FAO [Food and Agriculture Organization] (2011) Environment impact assessment: guidelines for FAO field project, 44 p

http://bali.bps.go.id/bali in figure 2013, November 26, 2013

http://bulelengkab.bps.go.id/buleleng in figure 2013, November 16, 2013

http://www.indonesia-investments.com, December 31, 2015

JICA (2006) The Comprehensive study on water resources development and management in Bali province, Republic of Indonesia. Public Works Agency, Bali Province. Denpasar. Rep, 254 p

Nakagiri T et al (2013) Basic diagnose of the present situation of water quality environment in the Saba river basin. In: presentation poster on 1st Bali Stakeholder Meeting on September 8, 2013, Collaboration between Research Institute for Humanity and Nature (RIHN) and Bogor Agricultural University, p 1

Regional Development Planning Agency of Bali Province (2012) Economic development plan in Bali province. In: presentation paper on Nature, Culture and Economic in Bali Island: international workshop, Nusa Dua Bali, September 5–6, 2012 in collaboration with Research Institute for Humanity and Nature (RIHN)-Japan, Udayana University, Bogor University of Agriculture, and Hasanuddin University

Setiawan BI et al (2013) Water availability and water productivity in the context of climate, land use and socio-economic changes in Saba and Jeneberang watershed, Indonesia. Collaboration between RIHN, IPB, Udayana University, and Hassanudin University. Bogor. Rep, 68p

Part IV
Practices in Chongming Island, China: Regional Management and Environmental Policy for Water Reclamation and Recycle

Chapter 8
Current State of Water Management in Chongming Island

Jianhua Li and Jun Nakajima

Abstract Chongming Island is the largest alluvial island in the world and is an ecologically sensitive area. It has 15,924 rivers with an overall length of 9352.3 km. Rapid economic growth in Chongming has presented great challenges to its water resource managers because of a scarcity of water resources, severe water pollution, growing domestic and industrial water demands, and the requirement for food security. This paper provides an overview of water resources and their management in Chongming. It describes the key water issues faced in Chongming, as well as the institutional, legal water regulatory arrangements, and the construction of an eco-island to address these challenges, including approaches to water resource allocation and management, pollution control, and water supply regulations. This paper concludes with a discussion of the priorities and challenges for water management, the progress that has been made, and the improvements that will be required to ensure the long-term sustainability of water resources in Chongming Island.

Keywords Water supply and demand • Water pollution control • Water resource dispatching • Sustainable management

8.1 Introduction

Chongming Island is located in the lower Yangtze Estuary, between 121°09′–121°54′E and 31°27′–31°51′N. Its current area is 1200 km^2. It is still increasing in size by about 500 ha annually through the deposition of sediment by the Yangtze River. At present, Chongming has two water sources: precipitation and the Yangtze River transit water resource. The annual water resource volumes are 35.37×10^8 m^3,

J. Li (✉)
College of Environmental Science and Engineering, Tongji University, Shanghai 200092, China
e-mail: leejianhua@tongji.edu.cn

J. Nakajima
College of Science and Engineering, Ritsumeikan University, Kusatsu, Shiga 525-8577, Japan

K. Nakagami et al. (eds.), *Sustainable Water Management*,
DOI 10.1007/978-981-10-1204-4_8

33.60×10^8 m^3, 32.15×10^8 m^3, and 31.05×10^8 m^3 at given guaranteed efficiencies of 20, 50, 75, and 90 %, respectively. Among them, the water resource volumes from the Yangtze River are 31.92×10^8 m^3, 30.15×10^8 m^3, 28.70×10^8 m^3, and 27.60×10^8 m^3, respectively. According to the Chinese government standard for water quality issued in 2002 (GB 3838-2002), the water quality is classified as category IV (i.e., a general industrial water zone and water recreation area where no direct contact with humans occurs). According to the water resource plan for Chongming Island, four new water supply plants have been completed by 2010, which increased the water supply volume by about 1.31×10^8 m^3 annually (Ni et al. 2012).

8.2 Current Characteristics of Water Resources in Chongming Island

8.2.1 Analysis of Water Resource Levels in Chongming Island

The water resource of Chongming Island is mainly composed of the local surface water resource and the transit water resource of the Yangtze River. The local surface water resource is runoff caused by rainfall. The average precipitation over many years is about 1050 mm, producing an annual average runoff volume of about 3.45×10^8 m^3. The annual average water diversion from the transit water resource of the Yangtze River is about 30.15×10^8 m^3. Among these resources, the local runoff volume accounts for 10.27 % and water diversion accounts for 89.73 %, an 8.74 ratio difference between the two values (Ma et al. 2008).

8.2.2 Analysis of Water Resource Quality in Chongming Island

According to the monthly water quality data from 25 monitoring sections between 2005 and 2008, comprehensive evaluation yields the following water resource quality results.

2005: Monitoring sections with class II water quality accounted for 4.5 % of the total, class III accounted for 40.2 %, class IV accounted for 42.6 %, class V accounted for 10.0 %, and water quality below class V accounted for 2.7 %. The qualified rate of river water quality was 44.7 % (above class III) and almost all of the water quality was in classes III to V. The main transnormal items were CODcr, TP, NH_3-N, and so on.

2006: Monitoring sections with class II water quality accounted for 8.0 % of the total, class III accounted for 56.5 %, class IV accounted for 27.7 %, class V accounted for 5.7 %, and water quality below class V accounted for 2.1 %. The qualified rate of river water quality was 64.5 % (above class III). Compared to 2005, the water quality was improved but still fell mostly within classes III to V.

The main transnormal items were CODcr, TP, NH_3-N, and so on. According to the monthly data, the water quality was lowest in July, when the qualified rate was only 35.7 %, and highest in August, when the qualified rate reached 85.7 %.

2007: Monitoring sections with class II water quality accounted for 6.0 % of the total, class III accounted for 49.1 %, class IV accounted for 33.9 %, class V accounted for 9.2 %, and water quality below class V accounted for 1.8 %. The qualified rate of river water quality was 55.1 % (above class III). Compared to 2006, the water quality was reduced and was again mostly within classes III to V. The main transnormal items were CODcr, TP, NH_3-N, and so on. According to the monthly data, the water quality was lowest in July, when the qualified rate was only 10.7 %, and highest in December, when the qualified rate reached 89.3 %.

2008: Monitoring sections with class II water quality accounted for 25.0 % of the total, class III accounted for 48.9 %, class IV accounted for 22.2 %, class V accounted for 2.1 %, and water quality below class V accounted for 1.8 %. The qualified rate of river water quality was 73.9 % (above class III). Compared to 2007, the water quality was improved and was mostly within classes II and III. The main transnormal items were CODcr, TP, NH_3-N, and so on. According to the monthly data, the water quality was lowest in February, when the qualified rate was only 42.9 %, and highest in December, when the qualified rate reached 92.0 %.

According to the monitoring of water quality in Chongming Island from 2005 to 2008, the river water quality is generally good; however, there are some areas where the water is of class V or worse. The major transnormal pollutants are CODcr, TP, and NH_3-N. Compared to the "Shanghai water (Environment) function regionalization," which indicates that the general water quality in Chongming Island should be class III, there is room for water quality improvement.

8.2.3 Analysis of Water Resource Levels in Yangtze Delta

Chongming Island is the world's largest alluvial island. Its coastal wetland and tidal flats provide many important ecological services (Zhao et al. 2004). Yangtze Delta has a half-day shallow tide (the tidal level rises and falls twice each day) and the average tidal cycle is 12 h 25 min. The deformation degree of the waveform is bigger when facing upstream, leading to changes in the tidal level, tidal range, and tidal time. The tidal level is higher when facing upstream; therefore, the tidal range progressively decreases when facing upstream, the tidal time duration from estuary to upstream shortens, and the duration of the tidal ebb lengthens. The phenomenon of unequal diurnal tides is quite strong. Annually, the highest tidal level usually appears during the typhoon, and the astronomical high tide and/or upstream flood occurs where the influence of the typhoon is relatively large (Liu et al. 2015).

Table 8.1 The historical characteristics of tidal level in Bao Town

Historical highest tide level (m)	Average highest tide level (m)	Historical lowest tide level (m)	Average lowest tide level (m)	Average tide level (m)	Rising tide duration	Falling tide duration
6.03	3.35	−0.19	0.89	2.12	4 h 38 min	7 h 48 min

Historical data indicate the characteristics of the tidal level station in Bao town (Table 8.1).

Yangtze Delta is a giant wet estuary with abundant water resources; the annual runoff of the main stream of the Yangtze River is 919.2 billion m^3.

8.2.4 Analysis of Water Resource Quality in Yangtze Delta

The Yangtze River estuary is the highest quality water source in Shanghai. We combined "The Environmental Impact Report of the Raw Water Project in Qingcaosha Water Source" and the monitoring water quality data for three points (Chong west, South gate, and Bao town) from 2004 to 2008 and performed an analysis of six indicators for Chongming heading south along the Yangtze River. The six indicators were dissolved oxygen (DO), permanganate index (COD_{Mn}), chemical oxygen demand (COD), five-day biochemical oxygen demand (BOD_5), ammonia (NH_3-N), and chlorides. The results are as follows.

1. Dissolved oxygen(DO)

The Yangtze River estuary is a huge water resource, with a complicated and fast water flow. Aeration is relatively extensive and pollution is relatively mild. The DO content of the Yangtze River estuary falls within classes I and II of "The Standard of Surface Water Environmental Quality" (GB3838-2002). The content of DO is influenced to a relatively large extent by water temperature; the content in winter is generally higher than that in summer. According to the monitoring data, the concentrations of DO in Chong West, South Gate, and Bao Town are mostly above 6 mg/L, reaching the class II water standard.

2. Permanganate index(COD_{Mn})

The permanganate index of each monitoring point is generally between 2 mg/L and 4 mg/L and its fluctuation is relatively low. Most values reach the class II water standard.

3. Chemical oxygen demand(COD)

The average COD of monitoring points in the Yangtze River estuary is below 15 mg/L. The values show large variation but reach the class II water standard. The highest concentrations of COD in Chong West, South Gate, and Bao town are below 20 mg/L, and they are mostly below 15 mg/L. The values reach the class II water standard.

4. Five-day biochemical oxygen demand (BOD_5)

The abundant DO creates conditions for the biochemical degradation of organic matter; therefore, the BOD_5 in the Yangtze River estuary is relatively low, generally below 3 mg/L, and meets the class II standard.

5. Ammonia (NH_3-N)

The concentrations of NH_3-N in the Yangtze River estuary are mostly within classes II and III.

6. Chlorides

The differences in chloride concentration between the various monitoring locations in the Yangtze River estuary are relatively large. Generally speaking, the closer to the estuary, the greater the influence of the outer seawater. According to the monitoring data of recent years, the content of chlorides in the Yangtze River estuary follows an increasing trend.

In conclusion, the water quality along the Yangtze River estuary of Chongming south is relatively good, reaching the class II and III standards. Changes in the water quality indicators are not large.

8.3 Challenges for Water Resource Management in Chongming Island

8.3.1 The Characteristics of Major Pollutant Sources in Chongming Island

8.3.1.1 Point Source Pollution

According to the results of an investigation into pollutant sources in Chongming Island in 2007, there are 233 industrial enterprises that emit sewage straight into the river. Annual sewage emissions are 5.6878 megatons, annual emissions of CODcr are 381.29 t, and annual emissions of NH_3-N are 29.38 t. The major pollutant source (according to the status of point source pollution in Chongming Island) was confirmed as COD_6. The total emissions of major point pollution sources whose annual emissions are not less than 2 t are 4.9759 megatons (87.48 % of the total), while such emissions of CODcr are 322.43 t (87.22 % of the total) and of NH_3-N are 23.25 t (82.81 % of the total). The distribution of the main polluting enterprises is shown in Fig. 8.1.

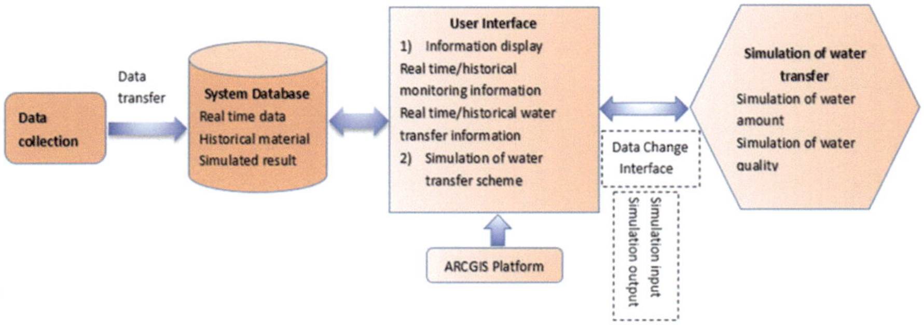

Fig. 8.1 System of water resource integrated regulation

8.3.1.2 Nonpoint Source Pollution

Nonpoint source pollution includes livestock, aquaculture, agricultural runoff, and domestic pollutant emissions. In 2007, the amount of CODcr produced by various types of nonpoint source pollution was 38775.06 t, and the amount of NH_3-N was 3293.02 t. The amount of CODcr produced by livestock breeding was 9265.25 t, and the amount of NH_3-N was 1853.05 t. The amount of CODcr produced by aquaculture was 13010.75 t, and the amount of NH_3-N was 175.44 t. The amount of CODcr produced by farmland surface runoff was 8199.85 t, and the amount of NH_3-N was 17,618 t. The amount of CODcr produced by domestic pollution sources was 8299.21 t, and the amount of NH_3-N was 1088.35 t. Regarding nonpoint source pollution in Chongming Island, a detailed study with a focus on spatial variation was carried out by Zhu et al. (2010) (Tables 8.2, 8.3, 8.4, 8.5, and 8.6).

1. Livestock breeding
2. Aquaculture
3. Surface runoff from farmland
4. Domestic pollution

8.3.2 The Security Requirement for Flood Control and Waterlogging Prevention

In 2008, the resident population of Chongming Island was 68 million. The whole county added a value of 13.77 billion yuan, an increase of 12.1 % over the previous year. It had a fiscal revenue of 5.07 billion yuan, an increase of 19.4 % over the previous year. As an important strategic region of Shanghai for sustainable development in the twenty-first century, building the ecological security of Chongming Island is necessary; in this regard, carrying out the security measures of flood control and waterlogging prevention is the first requirement.

Table 8.2 Pollutant discharge amounts from the raising of livestock and poultry in different townships of Chongming Island

Name of township	Equivalent of hogs	COD_{Cr} (t/a)	NH_3-H (t/a)	Amount of pork raised	Amount of poultry	Amount of laying hens	Amount of other poultry	Amount of dairy cows
Luhua	30,660	559.55	111.91	18,477	137,000	81,000	56,000	855
Xincun	14,839	270.82	54.16	14,056	25,000	22,000	3000	0
Sanxing	51,127	933.06	186.61	44,310	274,000	135,000	139,000	0
Miao	64,516	1177.42	235.48	54,503	337,000	184,000	153,000	133
Gangxi	22,869	417.36	83.47	20,969	82,000	32,000	50,000	0
Jianshe	31,252	570.35	114.07	20,702	272,000	166,000	106,000	325
Xinhe	27,010	492.93	98.59	19,743	189,000	109,000	80,000	230
Shuxin	34,029	621.04	124.21	27,696	250,000	130,000	120,000	0
Gangyan	28,786	525.34	105.07	23,969	218,000	71,000	147,000	0
Xianghua	21,103	385.14	77.03	19,670	59,000	27,000	32,000	0
Zhongxing	25,223	460.33	92.07	18,140	144,000	116,000	28,000	275
Chenjia	29,694	541.91	108.38	19,677	73,000	39,000	34,000	815
Chengqiao	24,500	447.13	89.43	21,117	35,000	18,000	17,000	250
Bao	34,088	622.11	124.42	31,305	94,000	73,000	21,000	0
Farm	67,987	1240.76	248.15	67,987	0	0	0	0
Total	507,683	9265.25	1853.05	422,321	2,189,000	1,203,000	986,000	2883

1. The amount of pork in large-scale hoggery accounts for 10 % of the total; there is no recycling
2. Sixty broiler chickens can be converted to one hog equivalent (one laying hen can be converted into two broiler chicken equivalents), and one beef cattle can be converted into five hog equivalents (one dairy cow can be converted into two beef cattle equivalents)
3. The COD_{Cr} emission coefficient of one hog is 50 g/day and the NH_3-H emission coefficient is 10 g/day

Table 8.3 Annual pollutant discharge amounts for aquaculture (fishponds) in different townships of Chongming Island

Name of township	Area of fishpond (ha)	COD_{Cr} (t/a)	NH_3-H (t/a)
Luhua	1025	1024.17	13.84
Xincun	1115	1114.28	15.06
Sanxing	559	558.74	7.55
Miao	938	936.96	12.66
Gangxi	400	399.10	4.99
Jianshe	370	369.33	4.99
Xinhe	1237	1235.36	16.69
Shuxin	513	512.19	6.92
Gangyan	636	635.76	8.59
Xianghua	69	68.53	0.93
Zhongxing	956	955.14	12.91
Chenjia	1299	1297.90	17.54
Chengqiao	619	618.08	8.35
Bao	298	297.20	4.02
Farm	2991	2988.01	40
Total	13,025	13010.75	175.44

Comment: The COD_{Cr} emission coefficient of aquaculture is 999 kg/ha.a and the NH_3-H emission coefficient is 13.5 kg/ha.a

Table 8.4 Pollutant discharge amounts of surface runoff for farmland in different townships of Chongming Island

Name of township	Area of cultivated land (ha)	COD_{Cr} (t/a)	NH_3-H (t/a)
Luhua	866.0	100.37	2.16
Xincun	1408.2	163.21	3.51
Sanxing	4419.6	512.23	11.00
Miao	5139.1	595.62	12.80
Gangxi	1739.9	201.65	4.33
Jianshe	2168.9	251.38	5.40
Xinhe	3489.1	404.39	8.69
Shuxin	3202.0	371.11	7.97
Gangyan	4829.5	559.74	12.03
Xianghua	2738.3	317.37	6.82
Zhongxing	2682.7	310.92	6.68
Chenjia	4209.0	487.82	10.48
Chengqiao	3010.7	348.94	7.50
Bao	3127.4	362.47	7.79
Farm	27719.0	3212.63	69.02
Total	70749.4	8199.85	176.18

Table 8.5 Domestic pollutant discharge amounts in different townships of Chongming Island

Name of township	Total population	City			Country			Total COD_{Cr} (t/a)	Total NH_3-H (t/a)
		Population	COD_{Cr} (t/a)	NH_3-H (t/a)	Population	COD_{Cr} (t/a)	NH_3-H (t/a)		
Luhua	9041	1415	41.32	3.10	7626	45.65	11.13	86.97	14.23
Xincun	11,930	1271	37.11	2.78	10,659	63.80	15.56	100.92	18.35
Sanxing	40,409	2766	80.77	6.06	37,643	225.33	54.96	306.10	61.02
Miao	60,350	9507	277.60	20.82	50,843	304.35	74.23	581.05	95.05
Gangxi	29,616	6652	194.24	14.57	22,964	137.46	33.53	331.70	48.10
Jianshe	32,273	6661	194.24	14.59	25,612	153.31	37.39	347.81	51.98
Xinhe	51,637	17,085	498.88	37.42	34,552	206.83	50.45	705.71	87.86
Shuxin	45,624	8281	241.81	18.14	37,343	223.54	54.52	465.34	72.66
Gangyan	53,841	9020	263.38	19.75	44,821	268.30	65.44	531.68	85.19
Xianghua	32,345	6765	197.54	14.82	25,580	153.12	37.35	350.66	52.16
Zhongxing	30,998	5453	159.23	11.94	25,545	152.91	37.30	312.14	49.24
Chenjia	62,211	13,780	402.38	30.18	48,431	289.91	70.71	692.28	100.89
Chengqiao	100,344	71,836	2097.61	157.32	28,508	170.65	41.62	268.26	198.94
Bao	64,407	29,261	854.42	64.08	35,146	210.38	51.31	1064.81	115.39
Farm	25,539	0	0	0	25,539	152.88	37.29	152.88	37.29
Total	650,565	189,753	5540.79	415.57	460,812	2758.42	672.79	8299.21	1088.35

Comment: The CODCr emission coefficient of a rural resident is 16.4 g/person.d and the NH_3-H emission coefficient is 4 g/person.d. The CODCr emission coefficient of an urban resident is 80 g/person.d and the NH_3-H emission coefficient is 6 g/person.d

Table 8.6 Summary of nonpoint pollution sources on Chongming Island

Number	Category	COD_{Cr} (t/a)	NH_3-N (t/a)
1	Livestock breeding	9265.25	1853.05
2	Aquaculture	13010.75	175.44
3	Surface runoff from farmland	8199.85	176.18
4	Domestic pollution	8299.21	1088.35
Sum		38775.06	3293.02

8.3.3 The Requirements for Water Source Supply Assurance

According to the current distribution (Table 8.7), there are 33 water plants (including seven farm water plants) in Chongming Island. According to "The Professional Planning Revision of Water Supply Systems in Chongming Island Domain" and "The Intensive Water Supply Planning in Shanghai Suburb," approximately 94 % of the raw water comes from the inland river and the other 6 % comes from the Yangtze River or groundwater. The drinking water resource of Chongming Island will gradually shift from the inland river to the Yangtze River, and the intake water mode will adjust from decentralized to centralized. Appropriate planning will allow for the formation of the Dongfeng Xisha reservoir water source and the "four plants four sides." During the transition period, part of the inland river of Chongming Island will be required to supply water resources to the relevant water plants.

8.3.4 The Water Demands for Production

Agriculture is the main economic pillar of Chongming Island. In 2008, the island's arable land covered 49982.6 ha and fruit forest covered 10,489 ha. The afforested area was 22,607 ha, accounting for about 65 % of the total area. Water use for agricultural production depends largely on the inland surface water. There are some companies on the island that take water directly from the nearby river. The use of water for aquaculture on the island also largely depends on the river water. In addition, the cyclic island river and parts of the vertical main rivers are channels (the highest grade of such channels is VI), which bear the shipping function. In order to ensure appropriate navigation safety and meet the shipping draft depth, water level changes of the inland river must comply with the control indicator requirements of the navigable water level. For this reason, rational management of water resources is required to meet the demands of production water supply and provide, as far as possible, adequate and improved quality water.

Table 8.7 List of current plants in Chongming Island

Category	Number	Name	Commissioning data	Scale (10,000 m^3/d)	Water supply scope	Location of source water
Country plants	1	Chenqiao	1973	1.00	Suburb	Yangtze river
	2	Laojiao	1986	3.00		Laoxiao river
	3	Baozhen	1980	2.00	Baozhen	South Heng
	4	Xinhe	1990	1.00	Xinhe	South Heng
Village Plants	5	Yuan	1982	0.50	Chenjiazhen and Yuan	Groundwater
	6	Chenjiazhen	1998	1.00	Chenjiazhen	Xijiagang
	7	Zhongxin	1988	0.86	Zhongxin town	South Heng
	8	Xianghua	1972	1.00	Xianghua	Liuxiaogang
	9	Wuxiao	1987	0.64	Wuxiao	South Heng
	10	Hexing	1980	0.80	Hexing	Sixiaohe
	11	Gangyan	1997	1.00	Gangyan	Baozhengang
	12	Baoxi	1998	0.30	Baoxi	South Heng
	13	Shuxin	1986	1.00	Shuxin	South Heng
	14	Daxin	1981	0.96	Daxin	Zhihegang
	15	Xinmin	1983	0.50	Xinhe	Xinhegang
	16	Datong	1986	0.72	Datong	Zhangwanggang
	17	Jianshe	1991	0.96	Jianshe	Laoxiao river
	18	Gangxi	1995	1.00	Gangxi town	Laoxiao river
	19	Houjiazhen	1976	0.96	Aoshan	Zhangwanggang
	20	Jiangkou	1980	0.74	Jiangkou	South Heng
	21	Miaozhen	1976	0.67	Miaozhen	Egegang
	22	Hezuo	1980	0.72	Hezuo	Gelonggang
	23	Haiqiao	1991	0.80	Haiqiao	Miaogang
	24	Sanxing	1984	0.90	Sanxing	Jie river
	25	Xincun	1981	0.70	Xincun	Jie river
	26	Lvhua	1984	0.50	Lvhua	Xinjian river

(continued)

Table 8.7 (continued)

Category	Number	Name	Commissioning data	Scale (10,000 m^3/d)	Water supply scope	Location of source water
Farm plants	27	Yuejin	1975	1.00	Yuejin	Yuejin river
	28	Xinhai	2004	0.96	Xinhai	North Heng
	29	Changjiang	1981	0.80	Groundwater	Xinhegang
	30	Dongfeng	1985	0.48	Groundwater	North Heng
	31	Changzheng	2004	0.96	Changzhen	Minshengang
	32	Qianjin	1985	0.50	Farm in Qianjin	North Heng
	33	Qianshao	1997	0.15	Farm in Qianshao	Ground water

8.3.5 Water Demands for the Ecological Environment

Combined with the establishment of an "ecological island" and a "Green Industry Demonstration Base," which are proposed by the master plan of Chongming Island, the "water functional zoning in Shanghai" established the surface water in Chongming Island as an ecological water supply (primarily a landscape recreational water supply); the standard of the water quality is class III.

The diversity of rivers and water functions set forward higher requirements for the condition of river network hydrodynamics, water environments, and water resources. The relevant departments steadily advance the work of sewage interception and treatment and the development of the routine work of diverting clean water operations. Such work assists in recovering the water mobility of the river network, improving the hydrodynamic conditions, enhancing the self-purification capacity, improving the water environment, and repairing water ecology.

8.4 Measurement of Water Resource Management in Chongming Island

8.4.1 Research on Water Resource Regulation Methods

In Chongming Island, drinking water of highest quality is found in the southern branch water area of the Yangtze River estuary. The regular water resource regulation "South diversion North discharge, West water dispatch to East" is reasonable and practicable by optimizing the control conditions of water sluices and the water level of water resource regulations. This is done by selecting the opening height of the sluice gates of water diversion and drainage, and the guideposts of the high and low water level, trimming part of water diversion and drainage methods. Increasing the amount of water diversion and drainage to various extents improves the water quality. After a multi-program comparison, the recommended regular water resource regulation scheme of Chongming Island is to "keep the current regulation method and optimize the sluice opening heights for water diversion and drainage." The average control water level is generally between 2.6 and 3.0 m. The highest control water levels of two water diversion sluices in Chong west are 3.0 m (Sanshahong flood sluice) and 3.2 m. The lowest control water level of the water drainage sluice is 2.6 m.

8.4.1.1 Details of Water Resource Reasonable Dispatching

During non-flood period we should ensure the dispatching of water resources using, improve the water quality dispatching; during flood period we should coordinate the safety dispatching of controlling flood, improve the water quality dispatching, and ensure using water dispatching. Under special circumstances, we should quickly respond to take special emergency dispatching.

8.4.1.2 Research on the System for Water Resource Integrated Regulation Management

The system structure of water resource integrated regulation management in Chongming Island is divided into four layers: a data collection/transfer layer, a data layer, a user interface, and a simulation system interface. The relationship between each layer is as follows.

8.4.2 Construction of an Eco-Island in Chongming Island

Chongming Island is the largest alluvial island in the world and is an ecologically sensitive area. It has been planned as a world famous eco-island for future development. For a small island to follow an integrated water management scheme, the development of an eco-island is an important concept within sustainable development.

8.4.2.1 Main Problems

Vulnerability of the Island

Some small islands are increasingly confronted with the classic contradiction between economic progress and environmental degradation, which usually shows as environmental, economic, and social vulnerability during the process of sustainable development. Like many other small islands, Chongming Island also faces these limitations. Natural characteristics, such as its small physical size, its ecological uniqueness and fragility, its limited terrestrial natural resource endowments, and its sensitivity to natural disasters, are considered the most important factors restricting the sustainable development of the island. A healthy natural ecosystem is the basis of all life support systems, including those of human well-being and social and economic development. How to overcome the natural vulnerabilities and improve the resistance and resilience of natural ecosystems is the main problem that should be solved during the process of eco-island construction. In this regard, we have taken a first step by carrying out relevant studies evaluating the river ecosystem services in Chongming Island (Jing and Li 2012).

Saltwater Intrusion

Saltwater intrusion is the most significant natural problem in Chongming Island. In dry seasons, rainfall in the upper reaches of the Yangtze River decreases, significantly decreasing runoff in the lower reaches. The push of tidewater and surface circulation results in the upstream movement of the salt tide in both the southern and northern branches of the Yangtze River. In addition, the salt tide in the northern branch usually mixes back into the southern branch; thus, the salt tide surrounds the island. In addition, water projects, such as dams, pumping stations, and reservoirs, along the upper regions of the Yangtze River largely reduce the runoff in the lower reaches and therefore increase saltwater intrusion. Most of the runoff of the Yangtze River passes through the southern branch. In humid seasons, freshwater from the upper river is abundant, and the multiyear average runoff of the southern branch in July and August reaches 47,000 m^3/s at the Datong water station. Freshwater diffuses to the North China Sea and the salinity in the estuary is very low. However, in the dry season, especially from January to March, the runoff of the Yangtze River markedly decreases; the average runoff is only 10,800 m^3/s. At this time, the salt tide surrounds the island. The maximal salinity in the estuary of the Yangtze River occurs during these dry seasons each year. According to data from the southern branch from 1992 to 2001 at the Baozhen, Nanmen, and Xinjian monitoring stations, the highest concentration of chloride at Baozhen station reached 5130 mg/L (salinity was 9.3 thousandths) on January 18, 1992; Nanmen reached 2730 mg/L (salinity was 5.0 thousandths) on March 25, 1999; and Xinjian reached 1780 mg/L (salinity was 3.2 thousandths) on March 21, 1999. The multiyear average concentration of chloride at Baozhen station from January to March was 491 mg/L, the Nanmen average was 244 mg/L, and the Xinjian average was 144 mg/L. According to the monitoring data from the northern branch from 1982 to 2001 at Baizha, the highest concentration of chloride was 14,900 mg/L (salinity was 26.9 thousandths) and the multiyear average concentration of chloride was 3749 mg/L (salinity was 6.8 thousandths) (Huang et al. 2008). The average content of chloride in the northern branch was far higher than in the southern branch. The northern branch was surrounded by salt tide for almost the entire year, and water in the river showed some properties of the inshore seawater. This branch has already lost its function as the freshwater source for the island. In addition, rising sea levels in the future will decrease the thickness of the freshwater lens; thus, the availability of freshwater, and an increase in the frequency and intensity of storms, will lead to increased incidences of soil and freshwater contamination by storm over-wash. Saltwater intrusion results in soil salinization and a potential freshwater crisis on the island. The Yangtze River is the most important freshwater source for the island. In dry seasons, some water dams on the island cannot effectively pump freshwater due to salt tide intrusion. In some cases, saltwater intrusion results in a freshwater crisis, which may hold back the social and economic development of the island.

8.4.2.2 Integrated Water System Management

An advantage of Chongming Island is its available freshwater resource from the southern branch of the Yangtze River. Rational freshwater resource utilization would improve the water quality and protect the sources from saltwater intrusion. The water system on the island is largely a man-made system. There are two trunk canals, the south diversion canal (77 km) and the north diversion canal (34 km), which run west to east; 31 subtrunk canals with a total length of 371 km, which run south to north; 420 township-level canals that connect the trunk and subtrunk canals with a total length of 1118 km; and 19,280 village-level canals with a total length of 9338 km. The average canal net density is about 9.4 km/km^2, and the water surface rate is about 9.4 %. There are 24 water supply or drainage dams around the island, which have a total water supply capacity of 3210 m^3/s and a water drainage capacity of 250 m^3/s (Huang et al. 2008). Dams are the controlling infrastructures that connect the man-made water system with the Yangtze River. Though these dams and canals have played an important role in ensuring the safety of the water supply on the island, there has never been an integrated trunk canal around the island. This has caused slow water renewal and decreasing water quality, and the canals have filled with silt, sediment, and waste. Two measures should be taken to improve the water renewal rate and water quality on the island. First, the canal system should be repaired and updated. In order to connect the south canal with the north canal, an integrated trunk canal around the island should be built to carry freshwater from dams along the south branch of the Yangtze River to the inner canals. The south and west segments of the trunk canal would serve as freshwater input segments, and the north segment of the trunk canal would serve as a drainage segment. Some control dams and man-made lakes should be built as buffers between the trunk canal and other small canals. The freshwater input from the Yangtze River that passes through each canal would also increase the water renewal rate of these canals and improve water quality. In addition, freshwater input to the canals would largely complement groundwater, thus preventing soil salinization. Scientific management of the canal system would contribute to the maintenance of water and soil quality. Second, natural riparian buffer zones should be established along the canals to prevent soil erosion and agricultural nonpoint pollution. Ecological engineering (such as artificial wetlands and bank protection work) in the buffer zones would improve their efficacy at protecting the water system. In conclusion, the integrated management of the water system should ensure both the freshwater supply and the ecological security of water bodies (Cheng and Hu 2012).

8.5 Conclusions

The assessment of the current state of water management in Chongming Island on the basis of statistical data has illustrated that Chongming Island water utilities are facing three main challenges: water scarcity and declining water quality, vulnerability of the technical base, and construction of an eco-island. Based on the current social, economic, and water resources conditions, Chongming Island should adopt demand management as the primary strategy, while controlling water pollution to quench the increasing water demands from the rapid social and economic growth. Important issues, such as water use efficiency, water rights and water rights trade, and effective enforcement of laws and regulations, should be addressed while switching to demand management. Meanwhile, control and prevention of water pollution to improve the overall quality of the freshwater resource is also a crucial component in water resource management. Chongming Island should develop a sustainable water resource management strategy, such as the construction of an eco-island, in order to ensure that the limited water supplies meet the demands of economic development, social well-being, and the conservation of ecosystems in the context of global climate change.

In conclusion, to ensure that this high-quality service continues, it should be stressed that water utilities and water authorities should implement certain measures and policies to strike a balance between economic development and the protection of water resources, so as to maximize the value of existing supplies while ensuring the long-term sustainability of their use. Planning is also essential for both water utilities and water authorities to identify the priorities and the timeframe for their implementation.

References

Cheng H, Hu Y (2012) Improving China's water resources management for better adaptation to climate change [J]. Clim Chang 112(2):253–282

Huang B, Ouyang Z, Zheng H et al (2008) Construction of an eco-island: a case study of Chongming Island, China [J]. Ocean Coast Manag 51(8):575–588

Jing Z, Li J. (2012) The evaluation of river ecosystem services in Chongming Island, China[C]/ Proceedings of the World Automation Congress (Puerto Vallarta, 24–28 June 2012). IEEE, pp 1–5

Liu XW, Xia YQ, Xu JJ (2015) Impacts of runoff and tides on offshore water level in south of Chongming Island[C]/Water Resources and Environment: Proceedings of the 2015 International Conference on Water Resources and Environment (Beijing, 25–28 July 2015). CRC Press, p 81

Ma Y, Zou C, Sun Q et al (2008) Prediction on ecological water demand in Chongming Island based on system dynamics [J]. Chin J Ecol 1:024

Ni X, Wu Y, Wu J et al (2012) Scenario analysis for sustainable development of Chongming Island: water resources sustainability [J]. Sci Total Environ 439:129–135

Zhao B, Kreuter U, Li B et al (2004) An ecosystem service value assessment of land-use change on Chongming Island, China [J]. Land Use Policy 21(2):139–148

Zhu Y, Li J, Jiang D (2010) Application of principal component analysis in evaluation of water quality of Chongming Island [J]. J Anhui Agric Sci 38(22):11906–11908, 11912

Chapter 9
The Characteristics of Eutrophication and Its Correlation with Algae in Chongming Island's Artificial River Network

Jianhua Li, Xiaochen Chen, Jia Niu, and Xiaofeng Sun

Abstract In order to study the trophic status of rivers in Chongming Island, the temporal and spatial variations of environmental factors and their interaction mechanisms in the main rivers were investigated. Results showed that eutrophication was present in the main rivers of Chongming Island. The concentrations of nitrogen and phosphorus in the main rivers were high, and nitrogen pollution was particularly severe. In fact, those nutrients mainly originated from the Yangtze River, rather than the local source in this island. Except for in winter, the chlorophyll-a content in most of the main rivers reached some level of eutrophication. The appearance of phosphorus promoted the growth of algae; this was due to water from the Yangtze River entering the main rivers. The evolution of the algae population accorded with the PEG model, as follows: winter (diatom), spring (chlorophyta and diatom), summer (cyanobacteria and chlorophyta), and fall (diatom). In addition, the trophic state index (TSI_M) indicated that rivers were at the "heavy eutrophication" level in spring and summer, and the "eutrophication" level in winter and autumn. In summer, chlorophyll-a and TP were the main contributors to the trophic status of the rivers.

Keywords Algae • Chongming Island • Environmental parameter • Eutrophication • TSI_M • Water quality

---

J. Li (✉) • X. Sun
College of Environmental Science and Engineering, Tongji University,
Shanghai 200092, China
e-mail: leejianhua@tongji.edu.cn

X. Chen
Ritsumeikan – Global Innovation Research Organization, Ritsumeikan University,
Kusatsu, Shiga 525-8577, Japan

J. Niu
Research Center for Sustainability Science, Ritsumeikan University,
Ibaraki, Osaka 567-8570, Japan

K. Nakagami et al. (eds.), *Sustainable Water Management*,
DOI 10.1007/978-981-10-1204-4_9

9.1 Introduction

Eutrophication of waterways has become a significant issue during the development of many countries. It is mainly presented as an overgrowth of algae related to nutrients, light, temperature, hydrological conditions, and other environmental factors. The most obvious indicator of eutrophication is the enhancement of algae biomass and a variation in species structure, whereas the distribution and succession of algae are impacted by the physiological characteristics of the algae and environmental factors (nutrient, light, and temperature). Therefore, it is important to study the interaction mechanisms between algae and environmental factors.

Chongming Island (E 121° 09′30″–121° 54′00″, N 31° 27′00″–31° 51′15″) is located in the estuary of the Yangtze River; it is surrounded by this river and the ocean. It is the third largest island in China and the largest river estuarine deposit island. It has a length of 80 km and an area of 1267 km^2. With flat topography, the elevation of more than 90 % of the land is between 3.21 and 4.20 m. The main cluster of cities comprises Chengjiao city, Baozhen city, Xinhe city, Miaozhen city, Shuxin city, Xianghua city, Zhongxin city, and Chenjia city. It is characterized by a high population density and a developed economy. Forests and farms are mainly distributed in the north of Chongming Island.

Chongming Island is influenced by the north subtropical monsoon and has a mild and moist climate. The annual temperature averages 15.3 °C, with monthly average temperatures reaching their lowest level in January (2.9 °C) and highest level in July (27.6 °C). There is an abundance of rain in Chongming Island. It has been shown that the annual average rainfall is 1025 mm, with the rainfall between May and September accounting for 61.9 % of the total.

Water resources in Chongming Island are formed by water diversion from the Yangtze River and surface runoff. The annual water supply consists of 3.64×10^8 m^3 of surface runoff and 30.15×10^8 m^3 of water diversion from the Yangtze River (Yin and Xu 2005). Due to the geographic characteristics of Chongming Island, the water resource is mainly controlled by sluice gates. According to the regulations of water conservancy, the brake is opened to sustain the water level at 2.9 m during periods of heavy farm irrigation and 2.8 m as the normal level. During typhoon events, the river water level drops to around 2.6 m.

Chongming Island is covered by a rich river network, with 109.42 km^2 of water (9.0 % of the total area of island). There are 33 municipal county watercourses with a total length of 390.22 km. The average water level is 2.6–3.0 m; this is affected by ecological water use, agricultural production, and domestic use. According to the overall planning targets and the characteristics of the water system, Chongming Island has a river network diversion system in the south and west of the island and drainage in the north and east of the island; this favors the integration of diversion, drainage, shipping, and irrigation.

9.2 Materials and Methods

9.2.1 Geographical Site

In this study, 57 sampling sites in Chongming Island were selected by using a global positioning system (GPS) according to a uniform grid (Table 9.1). From January 20 to October 18, 2010, field observations and sample collections were carried out (January 20, April 19, July 20, and October 18). Water samples were collected in polythene bottles at a depth of 0.5 m.

9.2.2 Analysis of Samples

The chemical and biological parameters of water quality (i.e., TN, TP, NH_3-N, NO_3^- N, and PO_4-P) were determined based on the standard methods of the China National EPA from the China National Environmental Monitoring Center. Environmental parameters (i.e., temperature, pH, DO, and conductivity) were monitored with a portable water quality analyzer (HACH sensION156, Germany), and Chl-a was measured with a Phyto-PAM (Walz, Germany) before the water samples were collected from the sites.

Table 9.1 The geographical distribution of sampling sites in Chongming Island

Name	Direction	Level	Serial number	Geographical site
Nanheng River	W-E	Municipal	NH-1~NH-15	121.1960E, 31.7581N~121.7716E, 31.4950N
Beiheng River	W-E	Municipal	BH-1~BH-14	121.1960E, 31.7581N~121.7736E, 31.5884N
Xinjiangang	S-N	Country	XJ-1~XJ-5	121.2114E, 31.7450N~121.2258E, 31.8236N
Gelonggang	S-N	Country	GL-1~GL-5	121.3402E, 31.6767N~121.4094E, 31.7672N
Laoxiaogang	S-N	Country	LY-1~LY-5	121.4170E, 31.6247N~121.4539E, 31.7358N
Xinhegang	S-N	Country	XH-1~XH-5	121.5191E, 31.5716N~121.5503E, 31.6938N
Baozhengang	S-N	Country	BZ-1~BZ-5	121.6071E, 31.5387N~121.6716E, 31.6197N
Baxiaogang	S-N	Country	BY-1~BY-5	121.7560E, 31.4871N~121.8175E, 31.5679N

9.2.3 *Statistic Analysis*

The trophic status of water mainly depends on the influence of environmental factors such as the nutrient and organic matter contents and transparency. To evaluate the trophic status of the Chongming River, the trophic state index (TSI_M) by Carlson was adopted (Ye 2010).

The computation formula for TSI_M is as follows:

$$TSI_M(Chl-a) = 10 \times \left(2.46 + \frac{Ln(Chl-a)}{Ln2.5}\right)$$

$$TSI_M(TP) = 10 \times \left(2.46 + \frac{(6.71 + 1.15 \times Ln(TP)}{Ln2.5}\right)$$

$$TSI_M(SD) = 10 \times \left(2.46 + \frac{(3.69 - 1.53 \times Ln(SD)}{Ln2.5}\right)$$

$$TSI_M(TN) = 10 \times \left(2.46 + \frac{(3.93 + 1.35 \times Ln(TN)}{Ln2.5}\right)$$

$$TSI_M(COD) = 10 \times \left(2.46 + \frac{(1.50 + 1.36 \times Ln(COD)}{Ln2.5}\right)$$

$$\begin{aligned} TSI_M = {} & W(COD) \times TSI_M(Chl-a) + W(SD) \times TSI_M(SD) \\ & + W(TP) \times TSI_M(TP) + W(TN) \times TSI_M(TN) + W(COD) \times TSI_M(COD) \end{aligned}$$

TSI_M, trophic state index; W(X), weight.

Evaluation rule:$TSI_M < 37$ poor nutrition; $37 \le TSI_M < 53$ moderate nutrition; $53 \le TSI_M < 65$ eutrophication; $TSI_M \ge 65$ heavy nutrition (Cai 1997; Cai and Hu 2006).

9.3 Results and Discussion

9.3.1 *Spatial and Temporal Variation of Physical Factors*

9.3.1.1 Temperature

Water temperature can directly affect the physical and chemical properties of water and the growth of aquatic organisms. Biological and chemical reactions in water have a close relationship with water temperature. Since Chongming Island is located in the subtropical monsoon region of East Asia, the intensity of solar radiation and water temperature have significant seasonal variations. The spatial and temporal variations of water temperature showed that the range of water temperature in winter was between 4.5 and 8.7 °C, with an average temperature of 6.6 °C. In spring, it was between 10.2 and 16.3 °C, with an average temperature of 12.0 °C; in summer, it was from 28.5 to 32.5 °C, with an average temperature of 29.8 °C; and in fall, it was from 20.2 to 24.7 °C, with an average temperature of 22.0 °C. However, there were no significant differences in water temperature among the sampling sites.

9.3.1.2 pH

pH is a basic indicator in environmental water chemistry research. Precipitation and the adsorption of elements in solution have a close relationship with pH. The variation of pH is mainly affected by CO_2 concentration, water temperature, biological activity, ionic strength, and other factors. The spatial and temporal variations in pH showed that the range of pH in winter was 7.56–8.2 (average 7.97), in spring was 7.45–8.2 (average 7.98), in summer was 7.46–8.82 (average 7.93), and in fall was 7.37–8.55 (average 7.88). Overall, pH was weakly alkali in Chongming Island and the seasonal variation was significant. Chongming Island exhibits a natural phenomenon where the salt tide of the south river rises and the salt tide of the north river flows backward; this is due to the influence of the tide in winter and spring and a reduction in runoff during the dry season of the Yangtze River. Some researchers have reported that the backward flow of the tide and seawater is the main reason for increased pH (Zhang et al. 2011). Therefore, in winter and spring, the high pH of Chongming River is mainly associated with the invasion of saline water from the Yangtze River. However, the pH in summer was significantly higher than in other areas and showed different seasonal variations in the middle of the Beiheng River and the south region of the Gelonggang River. Research showed that the outbreak of algae was severe. The range of chlorophyll-a concentration was 40.44–55.45 μg/L (average 45.98 μg/L). Algae outbreaks enhance photosynthesis, thereby increasing the ability of the algae to fix CO_2 (Mu et al. 2011). Thus, algae outbreaks caused a high consumption of CO_2 in the water, leading to the higher pH in summer.

9.3.1.3 DO

Dissolved oxygen (DO) is an important indicator that can reflect the status of water quality and the aquatic metabolic level. The spatial and temporal trends in DO showed that its concentration was higher in Chongming River and that it had a significant seasonal difference. The DO concentration was higher in winter and spring than in summer and fall. The range of DO concentration in winter was 8.01–11.59 mg/L (average 9.98 mg/L), in spring was 7.68–11.67 mg/L (average 9.67 mg/L), in summer was 4.22–12.89 mg/L (average 7.39 mg/L), and in fall was 3.96–10.88 mg/L (average 6.10 mg/L). The DO concentration had a close relationship with temperature, the strength of algae photosynthesis, and chemical oxidation. In winter and spring, high DO was mainly affected by low temperatures. Lower water temperature not only led to increased oxygen solubility in water but could also inhibit the consumption of chemical and biological oxygen. In addition to temperature, algae are another reason for the change in DO concentrations in the inland waterways (Zhang and Sun 2004). The chlorophyll-a concentration of the watercourse was significantly higher in summer than in the other seasons. The growth of algae increased photosynthesis intensity, which led to the enhancement of DO input; as a result, DO concentration also reached a higher level in summer.

9.3.1.4 Turbidity

The spatial and temporal variations in turbidity showed a concentration range in winter of 13.4–70.9 NTU (average 44.6 NTU), in spring of 18.7–80.3 NTU (average 43.7 NTU), in summer of 14.1–14.1 NTU (average 43.1 NTU), and in fall of 24.8–92.3 NTU (average 57.0 NTU). For municipal rivers, the turbidity of the Beiheng River decreased gradually from west to the east, and that of the Nanheng River showed a trend of first falling and then rising. The county rivers showed a gradually declining trend from south to north, as did the Gelonggang River. Water resources of Chongming Island mainly originate from the diversion of the Yangtze River and surface runoff. The intake from the Yangtze River accounts for 89.73 % of the total water quantity. The annual sediment load from the Yangtze River into the sea is 4.68×10^8 t, with 50 % deposited in the Yangtze River estuary (Liu 2006). Due to the influence of the tide, runoff, and wind, strong resuspension of sediment occurs in the Yangtze River estuary, resulting in abnormal water turbidity. Studies have shown that the suspended sediment concentration is up to 0.30 kg/m^3 in the surrounding estuary of Chongming Island and that seasonal variation of the sediment concentration is not obvious (Chen et al. 2004). Therefore, the high sediment concentration in the water diverted from the Yangtze River has an important influence on variations in turbidity at Chongming Island. The diversion from the Yangtze River is higher in fall than in winter and summer. The difference in diversion is the main reason that the turbidity concentration was higher in fall than in the other seasons.

9.3.2 Spatial and Temporal Variations in Nutrient Concentrations

9.3.2.1 TN

Consideration of total nitrogen (TN) in the main rivers of Chongming Island showed that in winter the range of values was 3.79–6.41 mg/L (average 4.99 mg/L), in spring was 3.41–5.18 mg/L (average 4.07 mg/L), in summer was 3.66–4.39 mg/L (average 4.04 mg/L), and in fall was 2.11–2.59 mg/L (average 2.33 mg/L). The difference in TN concentration of the interchannel in winter and spring (variation coefficients of 19.8 % and 19.8 %, respectively) was significantly higher than in summer and fall (variation coefficients of 6.4 % and 8.2 %, respectively). For municipal rivers, the TN concentration was higher in the Beiheng River than in the Nanheng River. For county rivers, the TN concentration was significantly higher in the western and central rivers (Xinjiangang River, Gelonggang River, Laoxiaogang River, and Xinhegang River) than that in the eastern rivers (Baozhengang River and Baxiaogang River). The spatial and temporal variations in TN concentration showed that in the Beiheng River and Nanheng River, the variation of TN concentration along the river was more significant in winter and spring than in summer and fall, showing a downward then rising trend. The peak in TN concentration appeared in the central region of Chongming Island (at the intersection of the Gelonggang River, Xinhegang River, and municipal river). Compared with the trend for the municipal rivers, the TN concentrations of the county rivers (in addition to Gelonggang River) were higher in the north than in the center and south, whereas the variation in TN concentration along the Gelonggang River was not significant across the four seasons.

9.3.2.2 TP

For the main rivers of Chongming Island, the range of total phosphorus (TP) concentrations was 0.17–0.22 mg/L (average 0.19 mg/L) in winter, 0.12–0.20 mg/L (average 0.14 mg/L) in spring, 0.20–0.25 mg/L (average 0.23 mg/L) in summer, and 0.11–0.18 mg/L (average 0.15 mg/L) in fall. The variation coefficients of TP for the four seasons were 10 %, 20 %, 6.9 %, and 16.7 %, respectively. For the municipal rivers, TP concentration was always higher in the Beiheng River than in the Nanheng River. For county rivers, spatial differences for the inter-watercourse did not change with season. In summer, the difference in TP concentrations was low. In other seasons, TP concentrations were significantly higher in western and central rivers (Xinjiangang River, Gelonggang River, Laoxiaogang River, and Xinhegang River) than in eastern rivers (Baozhengang River and Baxiaogang River). In the municipal rivers, variation in TP concentration along the rivers was not obvious at first. However, at the intersection between Gelonggang River, Xinhegang River, and municipal river, TP concentration showed a rising trend. For the county rivers, the trend in TP concentrations was similar to TN.

9.3.2.3 Ammonia-Nitrogen

For the main rivers of Chongming Island, we found that ammonia-nitrogen concentrations were in the range 0.7–2.33 mg/L (average 1.61 mg/L) in winter, 0.33–1.03 mg/L (average 0.64 mg/L) in spring, 0.32–0.72 mg/L (average 0.43 mg/L) in summer, and 0.07–0.27 mg/L (average 0.15 mg/L) in fall. The variation coefficients of ammonia-nitrogen in the four seasons were 36 %, 34.3 %, 32.6 %, and 40 %, respectively. For municipal rivers, the concentration of ammonia-nitrogen in the Beiheng River was always higher than in the Nanheng River. For county rivers, in winter and spring, ammonia-nitrogen concentrations were markedly higher in western and central rivers (Xinjiangang River, Gelonggang River, Laoxiaogang River, and Xinhegang River) than in eastern rivers (Baozhengang River and Baxiaogang River). The spatial and temporal variations in ammonia-nitrogen concentrations indicated that concentrations in the central municipal rivers were markedly higher than those of the west and east. The peak ammonia-nitrogen concentration appeared at the intersection of the Gelonggang River, Xinhegang River, and municipal river. For county rivers, the trend in ammonia-nitrogen concentrations was declining at first and then rising.

9.3.2.4 Nitrate-Nitrogen

For the main rivers of Chongming Island, the nitrate-nitrogen concentrations were in the range 2.61–3.81 mg/L (average 3.31 mg/L) in winter, 2.42–3.77 mg/L (average 3.08 mg/L) in spring, 2.85–3.45 mg/L (average 3.20 mg/L) in summer, and 1.76–1.96 mg/L (average 1.90 mg/L) in fall. The variation coefficients of the concentrations of nitrate-nitrogen in the four seasons were 14.1 %, 12.7 %, 6.6 %, and 3.2 %, respectively. For municipal rivers, the concentration was always higher in the Beiheng River than in the Nanheng River. For county rivers, in winter and spring, nitrate-nitrogen concentrations were significantly higher in Gelonggang River, Laoxiaogang River, and Xinhegang River than in the other rivers. The spatial and temporal variations showed a nitrate-nitrogen concentration trend that was similar to TN.

9.3.2.5 DOC

For the main rivers of Chongming Island, dissolved organic carbon (DOC) concentrations were 3.92–9.56 mg/L (average 6.83 mg/L) in winter, 3.47–7.25 mg/L (average 4.92 mg/L) in spring, 3.87–6.84 mg/L (average 5.74 mg/L) in summer, and 2.40–3.94 mg/L (average 3.12 mg/L) in fall. The DOC concentrations of the interchannel in winter and spring showed variation coefficients of 35.1 % and 23.8 %, respectively, values which were significantly higher than those in summer and autumn (variation coefficients of 18.8 % and 14.7 %, respectively). For the municipal rivers, DOC concentrations were higher in the Beiheng River than in the Nanheng River. For county rivers, in winter and spring, DOC concentrations were obviously higher in western and central rivers (Xinjiangang River, Gelonggang River, Laoxiaogang River, and Xinhegang River) than in eastern rivers (Baozhengang River and Baxiaogang River). The spatial and temporal variations in DOC concentrations showed similarities with the variation of nitrogen. Compared with the trend in municipal rivers, DOC concentrations of the county rivers in winter were higher in the north and south than in the central area (Xinjiangang River, Baozhengang River, Laoxiaogang River, and Xinhegang River). Variations in DOC concentrations along the Gelonggang River were not obvious. In the other seasons, DOC concentration gradually increased from the south to the north.

9.3.2.6 Chl-a

For the main rivers of Chongming Island, Chl-a concentrations were 5.22–13.35 μg/L (average 8.06 μg/L) in winter, 7.69–54.89 μg/L (average 21.76 μg/L) in spring, 14.87–41.36 μg/L (average 27.82 μg/L) in summer, and 6.26–34.07 μg/L (average 14.70 μg/L) in fall. According to the evaluation rule for eutrophication (Caspers H 1982), i.e., eutrophication is evident when the Chl-a concentration exceeds 11 μg/L, the results show that most rivers were in the "eutrophication" state. The variation coefficients of Chl-a in the four seasons were 32.1 %, 74.3 %, 31.3 %, and 60.4 %, respectively. The spatial and temporal variations in Chl-a concentration showed generally high concentrations, with 83.05 % of the sampling sites in the state of "eutrophication." For the municipal rivers, Chl-a concentrations increased rapidly after Beiying River entered the Chongming farm system and reached a peak at the intersection of the Gelonggang River, Xinhegang River, and Beiheng River. The peak Chl-a concentration of the Nanying River appeared at the intersection of the Nanheng River and Gelonggang River. The Chl-a concentrations of the country rivers showed a trend of increasing gradually from south to north (except for the Gelonggang River). In relation to the constitution of the algae population, diatom was the main species in winter, accounting for 66.01 % of total algae. In spring, green algae increased to 40.51 %, while diatom declined to 58.88 %, resulting in diatom and green algae becoming the dominant species. In summer, diatom dropped to 20.19 %, while the proportion of blue algae and green algae increased to 39.14 % and 40.67 %, respectively. In autumn, the proportions of blue algae, green algae, and diatom were 7.84 %, 19.35 %, and 72.81 %, respectively. On the basis of an analysis of the evolution laws for the planktonic algae of multiple temperate nutritional lakes, the PEG (Plankton Ecology Group) model was proposed (Sommer et al. 1986). The seasonal succession regularity of algae in PEG was: winter and spring, hidden algae and diatoms; summer, green algae; late summer or early autumn, cyanobacteria; and fall, algae. Our results show that the Chongming River had similar characteristics to those predicted by PEG; they were consistent with the evolution laws for algae.

9.3.3 Relevance of Chl-a and Environmental Factors

Through correlation analysis of the main physical and chemical indicators and chlorophyll-a (Table 9.2), it was found that water temperature and pH both had significant positive correlations with chlorophyll-a ($P<0.01$), while turbidity and

Table 9.2 Correlation analysis of environmental factors and Chl-a in the main rivers of Chongming Island

	Temperature	pH	DO	Turbidity	Chl-a	TN	TP	NH_3-N	NO_3^- N
pH	−0.16*	1							
DO	−0.73**	0.65**	1						
Turbidity	0.18**	−0.24**	−0.27**	1					
Chl-a	0.26**	0.22**	0.01	−0.37**	1				
TN	−0.51**	0.17*	0.36**	−0.44**	0.20**	1			
TP	0.10	0.08	0.02	−0.19	0.44**	0.41**	1		
NH_3-N	−0.68**	0.06	0.42**	−0.29**	−0.07	0.80**	0.31**	1	
NO_3^- N	−0.33**	0.15*	0.22**	−0.37**	0.18**	0.91**	0.32**	0.51**	1
DOC	−0.30**	0.06	0.20**	−0.36**	0.36**	0.71**	0.55**	0.64**	0.56**

Note: *significant correlation at 0.05 level, $n = 236$; **significant correlation at 0.01 level, $n = 236$

chlorophyll-a had a significant negative correlation ($P<0.01$). In terms of the nutrients (TN, TP, NH_3-N, and NO_3^- N) and DOC, there were significant positive correlations with chlorophyll-a ($P<0.01$), expect for NH_3-N.

Water temperature had different levels of influence on the metabolism of algae. Algae struggled to adapt to water temperature by changing their rate of cell division and production of Chl-a. In the appropriate temperature range, an increase in temperature can promote the growth of algae. The range of water temperatures at Chongming Island was 4.5–32.5 °C. A significant positive correlation between chlorophyll-a and temperature showed that within this range (4.5–32.5 °C), temperature can significantly promote the growth of algae.

The rapid growth of algae can reinforce its photosynthesis, yielding high pH and DO. This study showed that between chlorophyll-a and pH existed a positive correlation; therefore, the growth of algae can be considered the main reason for increased pH. Chlorophyll-a and DO did not show a significant correlation ($P>0.05$), suggesting that the activity of algae was not the main factor affecting the change in DO. In Chongming Island, high DO concentrations were due to low temperature conditions, which increased the solubility of DO ($P<0.01$).

In our study, turbidity was significantly negatively correlated with the concentration of chlorophyll-a because low turbidity causes increased transparency and thus promotes the photosynthesis of algae. Changes in turbidity were often accompanied by a change in hydrodynamics. Due to the influence of sediment, water diverted from the Yangtze River had high turbidity. This turbidity decreased after entering Chongming Island due to the effects of flow rate. The slower flow rate was of benefit to algae due to their absorption of nutrients, which helped to maintain the structure of the community (Lin and Han 2001; Zhao 2006). Hilton's study showed when water residence times are 2–3 times the doubling time of planktonic algae, such algae can grow fully (Hilton et al. 2006).

This study showed that TN and TP had significant positive correlations with chlorophyll-a ($P < 0.01$). In indoor cultivation experiments, Fang et al. (2006) found that under conditions of sufficient light, phosphorus was the main factor limiting algae growth in the Yangtze River estuary; our study shows the same result. The Redfield ratio (a ratio of nitrogen to phosphorus of 16) is considered to be the optimum ratio for healthy algae growth and physiological balance. In Chinese estuaries, the ratio of nitrogen to phosphorus ranges from 30:1 to 80:1; thus, available nitrogen is much greater than phosphorus, which leads to phosphorus being the main factor limiting algae growth. In Chongming Island, nitrogen pollution is relatively severe; the ratios of nitrogen to phosphorus in winter, spring, summer, and fall were 41.90, 58.25, 68.92, and 39.47, respectively. High nitrogen levels may also lead to TP becoming the main factor limiting the growth of algae.

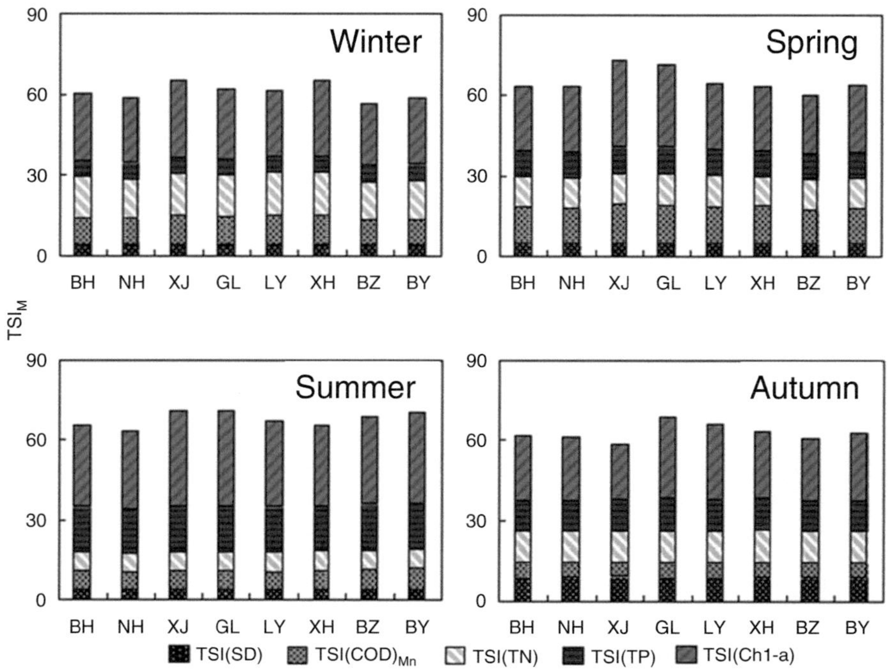

Fig. 9.1 Evaluation of the trophic states of the main rivers in Chongming Island in winter, spring, summer, and fall

9.3.4 Eutrophication Evaluation of Main Watercourses in Chongming Island

The evaluation of the trophic state of the main rivers in Chongming Island (Fig. 9.1) showed that the range of TSI_M values was 56.79–65.35 (average 61.12) in winter, 59.79–2.79 (average 65.15) in spring, 63.35–71.04 (average 67.90) in summer, and 68.81–68.88 (average 62.85) in fall. The variation coefficients of TSI_M in the four seasons were 5.07 %, 6.80 %, 4.39 %, and 5.15 %, respectively. The trophic state in spring and autumn reached the "heavy eutrophication" level, whereas in winter and autumn, it reached the "eutrophication" level. From the perspective of the contribution of various nutritional factors, the average contribution rate of TSI (chlorophyll-a) was the highest, reaching 41.93 %. The contribution rate of TSI (chlorophyll-a) in summer accounted for 47.71 %, which was higher than in the other seasons. The contribution rate of TSI (TP) in summer reached 25.67 %. The results suggest that chlorophyll-a and TP in summer were the main contributors to the trophic levels of the rivers. In winter, spring, and fall, the contribution rate of TSI (chlorophyll-a) declined to 6.38 %, 8.46, and 8.30 %, respectively; the contribution rate of TSI (TP) declined by 15.64 %, 11.12, and 15.64 %, respectively; and the contribution rate of TSI (TN) increased by 14.65 %, 14.65 %, and 7.12 %, respectively. It can be found that although Chongming River exhibited heavy nutrient levels all year round, the contribution of the main factors differed. In the summer, TSI (chlorophyll-a) and TSI (TP) were the major contributors to TSI_M. However, in the other seasons, TSI (TP) was not a main contributor to TSI_M. It is possible that blue and green algae were the dominant algae species in summer, both of which showed a positive correlation with TP; therefore, the contribution rate of TSI (TP) was higher. In the other seasons, diatom and green algae were the main species and the influence of TP on algae was lower; therefore, seasonal variation of the species structure was the main reason for changes in the contributory factors impacting the trophic states of the rivers.

References

Cai Q (1997) The review of lake eutrophication assessment methods. J Lake Sci 9:89–94

Cai Q, Hu Z (2006) Studies on eutrophication problem and control strategy in the three gorges reservoir. Acta Hydrobiol Sin 30:7–11

Caspers H (1982) OECD: eutrophication of waters: monitoring assessment and control. OECD Publication, Paris

Chen S, Zhang G, Yang S et al (2004) Temporal and spatial changes of suspended sediment concentration and resuspension in the Yangtze River estuary and its adjacent waters. J Geogr Sci 59:260–266

Fang T, Li D, Yu L et al (2006) Effects of irradiance and phosphate on growth of nanophytoplankton and picophytoplankton. Acta Ecol Sin 26:2783–2790

Hilton J, O'Hare M, Bowes MJ et al (2006) How green is my river? A new paradigm of eutrophication in rivers. Sci Total Environ 365:66–83

Lin Q, Han B (2001) Reservoir limnology and its application in water quality management: an overview. Chin J Appl Ecol 21:1034–1040

Liu J (2006) Research on dissolved inorganic nitrogen exchange at the sediment-water interface of the Yangtze estuarine tidal flats. East China Normal University, Shanghai

Mu C, Li K, Yu G (2011) Analysis of phytoplankton chlorophyll-a concentration in summer and autumn in Chaohu Lake and the impact factors of cyanobacterial blooms. J Biol 28:54–57

Sommer U, Gliwicz ZM, Lampert W et al (1986) The PEG-model of seasonal succession of planktonic events in fresh waters. Arch Hydrobiol 106:433–471

Ye L (2010) Studies on the eutrophication and spring phytoplankton bloom in Xiangxi Bay of Three-Gorge Reservoir. Graduate School of Chinese Academy of Sciences, Beijing

Yin G, Xu H (2005) Approach for the influence water system reconstructed to the water resources in Chongming Island. J Shanghai Constr Sci Technol 4:10–12

Zhang P, Sun J (2004) The influence of algae growing on pH and DO in surface water. J Environ Monit 20:49–50

Zhang Y, Sun J, Chen X et al (2011) Characteristics and preliminary analyses of the formation of pH in shallow groundwater in the Pearl River delta. J Hydrogeol Eng Geol 38:16–21

Zhao Y (2006) Study on effect of hydrometeorological element on algal growth. Hohai University, Nanjing

Chapter 10
Impacts of the Development on Land Use and the Water Environment

Ji Han and Xuepeng Qian

Abstract As the world's largest alluvial island and China's third largest island, Chongming enjoys significant geographical and ecological resources, which enables it to be at the forefront of Shanghai's future sustainable development. Ecological construction and environmental conservation are the two key development principles for the island's planning. With the support of a series of favored development strategies and policies, the process of urbanization in Chongming has been accelerating since the 2000s. This has triggered further changes in land use/cover and the local water environment. However, questions relating to (1) the extent to which the local environment has been affected by urbanization and (2) the policy implications of managing urbanization to achieve sustainable development have not yet been explicitly answered. Therefore, this chapter focuses on an assessment of the impact of urbanization on land use/cover change and the water environment by integrating statistics and remote sensing data. More importantly, certain policy deliberations related to ecological planning, water management, and the development of a low-carbon society are proposed to support the decision-maker in putting the eco-island plans into practice.

Keywords Eco-island planning • Land use/cover change • Water management • Sustainable development • Policy implications

10.1 Development Plan of Chongming Island

Chongming Island is the world's largest alluvial island and an important component of the Shanghai metropolitan region. It is located at the lowest point downstream of the Yangtze River between 121°09′–121°54′ East and 31°27′–31°51′ North.

J. Han (✉)
Shanghai Key Lab for Urban Ecological Processes and Eco-Restoration, East China Normal University, Shanghai, China
e-mail: jhan@re.ecnu.edu.cn

X. Qian
College of Asia Pacific Studies, Ritsumeikan Asia Pacific University, Beppu, Japan

K. Nakagami et al. (eds.), *Sustainable Water Management*,
DOI 10.1007/978-981-10-1204-4_10

It consists of three major islands – Chongming, Changxing, and Hengsha – and covers an area over 1200 km^2. It is expanding by approximately 500 ha per year due to the deposition of waterborne sediments from the Yangtze River (Ni et al. 2012). As the local climate is subtropical and oceanic, Chongming enjoys moderate temperatures (annual average, 15 °C), relatively high precipitation (annual average, 1025 mm), and four distinct seasons per year.

Due to its relatively isolated location away from mainland Shanghai, the exchange of the local population, materials, and information with the outside world has lagged far behind, meaning that the region's overall competitiveness does not match that of Shanghai. On the other hand, because of this limited economic development, Chongming has remained unexploited. Thus, it has retained significant geographical and ecological resources, enabling it to remain at the forefront of Shanghai's future sustainable development.

In terms of eco-environmental considerations, Chongming is characterized by agroecosystems and natural wetlands. Agroecosystems dominate the land use of the entire island; hence, Chongming has become one of the most important suppliers of agricultural products for the people of Shanghai. Natural wetlands widely exist on the island, especially along the coastline. These estuarine and coastal wetlands provide valuable ecosystem services, such as shoreline defense, water conservation and purification, fishery resources, recreational sites, etc. Of all the natural wetlands, Dongtan (Eastern Beach) wetland draws the most worldwide attention. It is a fertile area with a large number of aquatic animals of high economic value. It also functions as an indispensable staging and wintering habitat for Asia–Australia migratory birds as well as an important spawning and feeding ground for 63 species of fish. In 2002, Chongming Dongtan Nature Reserve was officially adopted into the Ramsar Convention as Wetlands of International Importance (rsis.ramsar.org/ris/1144).

Regarding the socioeconomic conditions of Chongming Island, it has remained at a primary stage of urbanization for some time, with a relatively low population density. Also, industries that are dominated by agriculture and secondary and tertiary industries have not left a significant mark. Overall, the local people have not significantly profited from the booming economy of Shanghai (Yuan et al. 2003; Wang et al. 2005). Geographically, Chongming Island is surrounded and isolated by the Yangtze River. This isolation explains its relatively slow development in the past. This isolation was mitigated in 2009 when the Shanghai Yangtze River Bridge (including a bridge and tunnel) was completed. Due to this improvement in accessibility, a considerable number of young residents now live or work in Shanghai city where they can earn a better life; this has, to some extent, resulted in a loss of the local labor force. Currently, the population on Chongming Island is about 700,000, approximately 75 % of which is involved in the agricultural sector (Huang et al. 2008). As the last "clean" space in the Yangtze River Delta, tourism is another source of income for Chongming Island, one that is becoming increasingly significant.

The development process of Chongming was once headed in an ill-advised direction. Up until the end of the twentieth century, there was no effective master plan proposed by either the local government or Shanghai municipal government. Certain inappropriate actions and behaviors drastically degraded the ecological value of Chongming Island (Zhao et al. 2004; Huang et al. 2008). Fortunately, the advent of the twenty-first century yielded a turning point. Complying with the Chinese central government's instructions to develop Chongming Island into a national ecological demonstration zone, Shanghai municipal government adopted some advanced ideas and released a master plan for the whole island in 2005. This determined to develop Chongming as an "eco-island," with an ultimate goal of developing beautiful landscapes, self-contained urban functions, a sustained economy, and a civilized society.

Accordingly, a series of packaged plans has been formulated and implemented to achieve a sustainable future for Chongming Island. To realize these targets, certain necessary processes have been put forward; these are divided into three stages as follows:

1. Near-term plan (to establish a tentative framework for eco-island construction before 2005). In detail, to strengthen the ecological construction and environmental protection so that important ecological function areas receive effective protection, to adjust the industrial structure transition toward a green economy, and to increase the construction of infrastructure so as to improve the quality of life of local people
2. Midterm plan (to achieve the development shape of the eco-island before 2010). To give full consideration to the advantages of the ecological resources, accelerate the adjustment of industrial structures, and greatly improve local people's well-being
3. Long-term plan (to achieve the eco-island before 2020). To complete the transition from pattern development to function development and realize the integration of production, life, and an eco-environment that includes a harmonious environment, diverse biological species, and a rich landscape

For the detailed ecological plan of the island, Chongming is divided into six ecologically functional areas and a natural reservation district, as described below:

- Functional area 1: west water ecological tourism and harbor logistics service area. This is targeted as an economic and tourism center, with water tourism and harbor logistics services as its supporting industries.
- Functional area 2: mid-west ecological agricultural area. This is developed as an important ecological agricultural area due to its strong agricultural foundation. Here, modern ecological agriculture will be the major supporting industry.
- Functional area 3: mid-north forest ecological tourism. This area is full of abundant natural resources, which makes forest ecological tourism possible. Ecological agriculture and tourism will be the supporting industries.
- Functional area 4: mid-south urban area, ecological industrial park. This area is expected to provide modern urbanization, green industry, and a comfortable eco-environment for the whole island and Shanghai. Accordingly, urban construction, a green industrial park, and tertiary industries will be developed as the dominant industries.

- Functional area 5: mid-east ecological agriculture area. This is planned as an ecological agricultural park. Green industry and agriculture, together with small town construction, will be the pillar industries in this area.
- Functional area 6: east harbor logistics service area, wetland ecological tourism area. Due to a strong economic foundation and abundant natural landscapes, harbor logistics services and ecological tourism will be promoted in this area.
- Functional area 7: east beach natural conservation area. In order to maintain the natural and seminatural ecological landscapes, this will be protected as an important natural conservation and ecologically sensitive area.

Overall, completion of the cross-sea Shanghai–Yangtze River Bridge has brought a new development era for Chongming Island. Ecological construction and environmental conservation are the two key development principles in the island's planning. A livable eco-environment and sustainable development will be the premises on which Chongming will become an international eco-island adjacent to a mega city. Due to these factors, and with the support of a series of favored development strategies and policies, the process of urbanization in Chongming has been accelerating since the 2000s. This has also triggered increased changes in land use/cover and the local socio-ecological environment including water, soil, air, biodiversity, human well-being, etc. To date, the eco-island plans have not been proposed in a very detailed manner. Instead, they showed promising directions for development. Questions related to (1) the extent to which the local environment has been affected by urbanization and (2) the policy implications of managing the quality of urbanization so as to achieve sustainable development have not yet been explicitly answered. Therefore, in this chapter, we assess the impact of urbanization on land use/cover changes and the water environment in Chongming by integrating statistics with remote sensing data. More importantly we propose some policy deliberations from the perspectives of ecological planning, water management, and the development of a low-carbon society to support the decision-maker in putting the eco-island dream into practice.

10.2 Land Use/Cover Change from 1990 to 2013

In order to understand the long-term impact of urban development on land use and cover in Chongming, especially before and after the implementation of eco-island planning, we use three remote sensing images as a basis for analyzing the spatio-temporal characteristics of land use/cover transition and landscape evolution from 1990 to 2013.

Three Landsat TM/ETM+ images taken in 1990, 2000, and 2013 were used in this study. Their spatial resolution was 30 m × 30 m. First, these images were preprocessed by The environment for visualizing images (ENVI) software and commonly used remote sensing techniques such as geometric correction and image stitching. Second, using a man–machine interactive interpretation method, the land use and vegetation coverage in Chongming were classified into six categories: cropland, forest, grassland, water, urban, and unused. Through a visual inspection method, we calculated that the overall accuracy was 85.1 % for 1990, 87.2 % for 2000, and 88.1 % for 2013, which is good enough to meet the minimum standard of 85 % proposed by the United States Geological Survey (Anderson et al. 1976). Third, we also chose three landscape indices to reflect the changes in landscape. These were (1) the Landscape Shape Index (LSI), which reflects the discreteness of landscape patches (the larger the LSI value, the more dispersed is a landscape); (2) the percentage of like adjacencies (PLADJ), which is an absolute measure of the aggregation of an urban landscape; and (3) the aggregation index (AI), which is calculated as an area-weighted mean class aggregation index. Equations for these three commonly used indices are given below.

$$\mathrm{LSI} = \frac{0.25\sum_{k=1}^{m} e_{ik}}{\sqrt{TA}} \tag{10.1}$$

$$\mathrm{PLADJ} = \left(\frac{\sum_{i=1}^{m} g_{ii}}{\sum_{i=1}^{m} \sum_{k=1}^{n} g_{ik}} \right) (100) \tag{10.2}$$

$$\mathrm{AI} = \left[\frac{g_{ii}}{\max\left(g_{ii}\right)} \right] (100) \tag{10.3}$$

For this analysis, Fig. 10.1 shows the land use/cover and landscape index changes in Chongming from 1990 to 2013. It is found that the urbanized area (including rural residential land) has grown significantly, while the area of cropland has decreased. Moreover, according to the changes in the landscape indices, both PLADJ and AI continually decreased during the entire period, which indicates that the urban land development has followed a trajectory of rapid sprawl rather than compactness. This is not a phenomenon unique to Chongming; it is common in most urbanized regions in China, as characterized by the rapid expansion of urban land at the expense of cropland. The local government has commonly expropriated rural land for urban development and forced rural residents to live in peri-urban or urban areas without providing them sufficient compensation and social services. This yields rapid urbanization at a low cost. However, due to a lack of supporting systems to help rural migrants to integrate into urban life, most choose to build houses back in their home village, usually on arable land. This acceleration of urbanization has caused a number of socio-environmental problems such as low-efficiency land use, decreased food security, and urban crime. Regarding the changes in LSI values, an initial increase was seen from 1990 to 2000 followed by a decrease from 2000 to 2013. This may reflect the influence of the urban development plan on the local landscape and environment. Before the implementation of the Chongming development plan in 2005, the landscape tended to be more dispersed due to the lack of overall planning. However, after 2005, with more attention being paid to

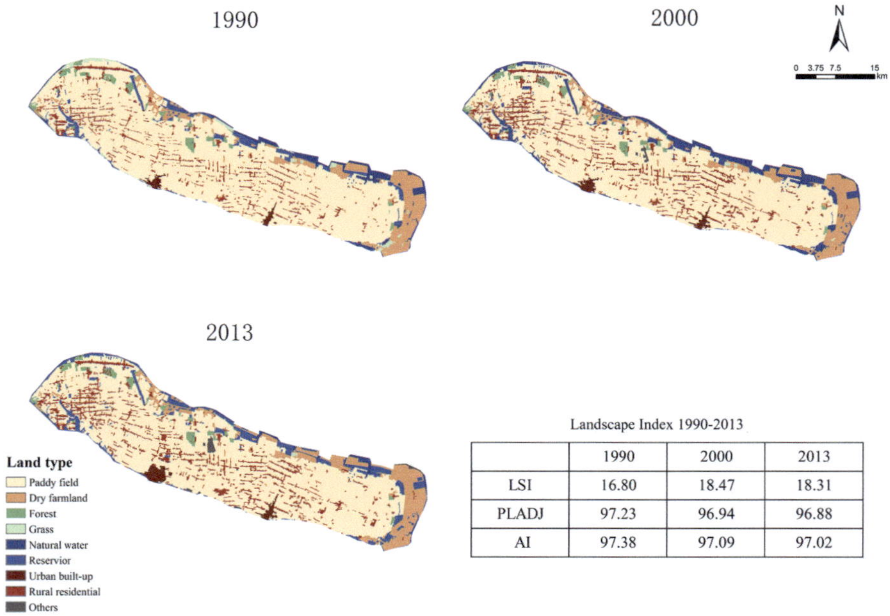

Landscape Index 1990-2013

	1990	2000	2013
LSI	16.80	18.47	18.31
PLADJ	97.23	96.94	96.88
AI	97.38	97.09	97.02

Fig. 10.1 Land use/cover transition matrices during 1990–2000 and 2000–2013 in Chongming

the eco-island planning in Chongming, the overall landscape has become more homogeneous.

To better understand the transformation of different types of land, we calculated the land use/cover transition matrices during 1990–2000 and 2000–2013. The results are shown in Table 10.1. Generally, the land use/cover change in Chongming was not as dramatic as in mainland Shanghai. It can be seen that major changes took place, i.e., the expansion of urban buildup, rural residential land, and water areas (especially reservoirs). On the one hand, the built-up area expanded greatly, and its encroachment on paddy fields accelerated during the last two decades (as aforementioned, this has been a common approach to increasing urban buildup because the expropriation of cropland in China is low in cost). On the other hand, the rural residential area, which is also considered a component of the urbanized area, has been undergoing rapid growth, with 32.5 km^2 of such land converted from paddy fields between 1990 and 2000 and the albeit slower growth of 7.0 km^2 converted from paddy fields between 2000 and 2013. This may be explained by the implementation of eco-island planning in 2005, with which the local government has placed much more emphasis on cropland protection. Moreover, the increased area of reservoirs and ponds represents another significant land use change; such areas were mainly converted from grass, paddy fields, and natural water bodies. Because of its prominent geographical location, Chongming contains one of the most important reservoirs for supplying drinking water to the whole of the Shanghai metropolitan area. It is also an important resource for the fishery industry in Shanghai.

Table 10.1 Land use/cover transition matrices in 1990–2000 and 2000–2013 (Unit: km^2)

Land type	Paddy field	Dry farmland	Forest	Grass	Natural water	Reservoir	Urban built-up	Rural residential	Others	Total
1990–2000										
Paddy field	803.88	0.12	0.05	0.00	0.36	4.66	2.59	32.49	2.15	846.29
Dry farmland	0.06	110.74	0.00	0.00	0.01	1.71	0.00	0.21	0.00	112.74
Forest	0.04	0.01	22.83	0.00	0.00	0.16	0.00	0.11	0.00	23.14
Grass	0.01	0.01	0.00	1.54	0.65	13.62	0.40	0.11	0.00	16.33
Natural water	0.01	0.02	0.00	0.00	31.05	12.61	0.00	0.24	0.27	44.21
Reservoir	0.04	0.02	0.00	0.00	0.00	45.08	0.00	0.00	0.00	45.14
Urban built-up	0.02	0.00	0.00	0.00	0.00	0.00	8.50	0.00	0.00	8.53
Rural residential	0.82	0.01	0.01	0.00	0.00	0.01	0.00	67.75	0.00	68.60
Others	0.00	0.00	0.00	0.00	0.00	0.00	0.00	0.00	0.05	0.05
Total	804.87	110.92	22.89	1.55	32.08	77.84	11.50	100.91	2.47	1165.03
2000–2013										
Paddy field	789.48	0.01	0.00	0.00	0.00	0.62	4.68	7.04	3.02	804.86
Dry farmland	0.01	107.64	0.00	0.00	0.77	1.98	0.00	0.52	0.01	110.92
Forest	0.00	0.00	22.77	0.00	0.00	0.00	0.00	0.12	0.00	22.89
Grass	0.00	0.62	0.00	0.79	0.00	0.14	0.00	0.00	0.00	1.55
Natural water	0.01	9.85	0.00	0.00	18.84	2.17	0.00	0.20	1.01	32.06
Reservoir	0.01	3.90	0.00	0.25	0.01	73.61	0.00	0.06	0.00	77.85
Urban built-up	0.01	0.00	0.00	0.00	0.00	0.00	11.49	0.00	0.00	11.50
Rural residential	0.15	0.00	0.00	0.00	0.00	0.00	0.00	100.76	0.00	100.91
Others	0.00	0.00	0.00	0.00	0.00	0.00	0.00	0.00	2.47	2.47
Total	789.67	122.02	22.78	1.04	19.62	78.51	16.18	108.69	6.49	1165.01

10.3 Water Environment and Management

Due to the eco-island master plan, the manufacturing industry has been declining under strict constraints. Hence, the domestic/commercial and agricultural sectors are the most water-demanding in Chongming Island. There are currently more than 10,000 waterways that intake and deliver water from the Yangtze River to Chongming Island. Since the water flow of the Yangtze River is diverted, Chongming Island is rich in water quantity. However, because of saline water intrusion and pollutant discharge, water quality has been seriously influenced. Nevertheless, this has been largely alleviated since March 2014 because Dongfeng Xisha Reservoir, located in the southwest of Chongming Island, has since been used as a stable water supply for Chongming.

Regarding the current status of the potable water supply in Chongming Island, the coverage rate is almost 100 %. There are two major water purification plants (Chenqiao and Chenjiazhen) and more than 30 medium-sized water purification stations. By 2015, all small water purification plants will be permanently shut down, and another two major water purification plants (Buzhen and Chongxi) will be completed. The total capacity of the water supply will reach 200,000 m^3/day, which is sufficient for the entire population of Chongming Island. At the same time, 75 % of the distribution pipes in the central area are to be replaced. As these were originally constructed 20 years ago, replacement and renewal is supposed to assure water quality and solve leakage problems. In the near future, the concentrated water supply system, which consists of one large reservoir and four major purification plants, will guarantee drinking water security for domestic and commercial use in Chongming.

Around 83 % of the total wastewater is currently treated by the 4 centralized wastewater treatment plants (Chenqiao, Xinhe, Buzhen, and Chenjiazhen), 11 medium-sized treatment stations, and several simple treatment facilities. This means that around one-fifth of the 150,000 tons of wastewater generated per day is discharged without sound treatment and management. In addition, due to a lack of technical and financial capabilities, operation and maintenance procedures are not sufficient. Moreover, the fluidity of many waterways in rural areas is low, which also contributes to poor water quality. Apparently, the introduction and upgrading of water treatment facilities and systems are essential for conserving and restoring a desirable water environment in Chongming Island. In the 12th 5-year Plan of Chongming Water Affairs Bureau, the coverage rate of the centralized sewer system should reach 80 % by 2015. However, if this percentage needs to be further increased, decentralized sanitation systems should be taken into consideration. For example, septic tanks and constructed wetlands are considered cost-efficient alternatives for remote rural areas. Several upscale and low-density resort housing zones are to be constructed away from central areas, where Johkasou, one of the most popular Japanese waste treatment technologies, could be adopted. According to the master plan, the land use of Chongming Island will be diverse. Therefore, it is important to introduce appropriate wastewater treatment systems that combine various technologies.

Building a low-carbon society is one of the most important topics in the eco-island master plan; this will also require energy-saving measures in the water utilization system. Reducing water demand and wastewater volume, as well as achieving an efficient utilization of the water resource, is consistent with the energy-saving requirement. Firstly, it is recommended to introduce rainwater utilization systems and water-saving equipment. The rainwater harvesting facilities are suitable for public facilities, while water-saving equipment can be applied to the new houses. Secondly, water reclamation and recycling may be another alternative for reducing water use demand. Wastewater condition, running costs, and energy consumption must be considered when designing such water reclamation systems because advanced treatment technologies, such as a membrane bioreactor (MBR), can be embedded. Additionally, a combination of the technologies and systems mentioned above may provide more effective and efficient solutions to save both water and energy. Since it is sometimes reported that local residents are dissatisfied with the cost burden related to water treatment, environmental education to increase resident knowledge, consciousness, and cooperation would also be indispensable.

Agriculture has always been the foundational industry in Chongming Island. The extent of the demand for water by agriculture is not clear because the irrigation water is drawn directly from the waterways by around 2000 pumping stations. One potential method to estimate this demand is to determine the water footprint based on the types of agricultural crops, the land area, and production. Investigation of the impacts of fertilizers and pesticides on the public water environment should also be conducted. According to the master plan, intensive agriculture will continue to be promoted. Those types of agriculture with high added value, such as organic farming, are bound to increase. This will bring new challenges to agricultural water management. Proper water treatment, such as using constructed wetland technologies, will be needed to reduce losses and the risk of agricultural nonpoint source pollution.

10.4 Development of Chongming Toward a Low-Carbon Society

Certain human activities, especially the anthropogenic emission of greenhouse gases (GHGs) such as carbon dioxide (CO_2), are considered major drivers of "global warming" in the modern era. Some environmental issues have already occurred due to this temperature increase. If this trend continues, ecocatastrophe may spread across the world (IPCC 2013). In an attempt to halt this trend, the concept of a "low-carbon society" has been put forward, which refers to a society with minimized GHG emission. The society combats global warming, and through conserving and restoring the environment, it can harmoniously coexist with nature (Ministry of the Environment Japan 2007).

To develop Chongming as a "low-carbon" eco-island, comprehensive strategies must be proposed. As one of the pioneering countries seeking a route to a "low-carbon society," Japan has an abundance of precious experience that could be shared and used as a reference. A decade ago, Japan launched its "Low-carbon Society" project, with the explicit goal of reducing CO_2 emissions to 70 % of their 1990 levels by the year 2050 (NIES 2008). Subsequently, a number of actions that could make substantial contributions to this effort have been put forward (see Table 10.2).

When it comes to putting the "low-carbon" concept into practice, almost all of these rational and effective actions could, with consideration of the specific local circumstances, be adopted by Chongming Island. In fact, some actions have already been taken. Several examples of the current achievements and further development proposals for Chongming Island are described below.

In terms of the commercial and residential sector, an "eco-office" demonstration building was designed and constructed in Chenjia Town, Chongming Island, by the joint efforts of Shanghai Chenjia Town Construction Development Co., Ltd. and Shanghai Research Institute of Building Sciences (Group) Co., Ltd. In 2013, the "National Green Building Innovation Award (first prize)" was granted to acknowledge all the green ideas and technologies involved in this building, including the use of green construction materials, natural ventilation and lighting, renewable energy, wastewater reclamation and reuse, eco-planting, low-energy consumption, intelligent control systems, and so on. In the future, the population density of Chongming Island is expected to increase, especially in its conventional residential community and its proposed upscale communities designed for business, leisure, and retirement. Hence, this "green building" movement should continue to be promoted, so as to minimize the environmental impact of such buildings throughout their life cycles.

From the perspective of transportation, the Shanghai municipal government and the local government of Chongming County should explore ways to effectively reduce dependence on private automobiles that burn fossil fuel. One suggestion for urban planning is to establish more compact communities, i.e., shortening the distance between apartment buildings so that residents can reach the nearest marketing center and public transportation station within a few minutes' walk or bicycle ride. In addition, it is reported that a new rail line will be constructed to connect Chongming Island directly to the downtown area of Shanghai city; this will also further reduce the use of private cars. Furthermore, integrated technological and management strategies/regulations could be formulated and implemented in a compulsory manner. In this way, conventional fossil fuels could be efficiently replaced by clean power. For example, the government could enforce the replacement of all conventional buses with new types using pollution-free fuel cells. Moreover, residents could be urged to use hydrogen-powered or motor-driven cars; otherwise, they must park their (conventional) cars outside the boundary of Chongming Island (Normile 2008).

With regard to the energy conversion sector, local renewable energy resources must be fully exploited (instead of conventional fossil fuels). For Chongming Island, four types of promising renewable energy must be emphasized: (1) solar radiation

Table 10.2 Actions contributing toward a low-carbon society

Action		General description	Sector
1	Comfortable and green built environment	Efficient use of sunlight and energy efficient built environment design. Intelligent buildings	Commercial and residential sector
2	Anytime, anywhere appropriate appliances	Use of top-runner and appropriate appliances. Initial cost reduction by rent and release system resulting in improved availability	
3	Promoting seasonal local food	Supply of seasonal and safe low-carbon local foods for local cuisine	Industrial sector
4	Sustainable building materials	Using local and renewable building materials and products	
5	Environmentally enlightened business and industry	Businesses aimed at creating and operating in low-carbon markets. Supplying low-carbon and high value-added goods and services through energy-efficient production systems	
6	Swift and smooth logistics	Seamless networking logistics systems with supply chain management, using both transportation and information and communication technology (ICT) infrastructure	Transportation sector
7	Pedestrian-friendly city design	City design requiring short trips and pedestrian (and bicycle)-friendly transport, augmented by efficient public transport	
8	Low-carbon electricity	Supplying low-carbon electricity by large-scale renewables, nuclear power, and carbon capture and storage (CCS)-equipped fossil (and biomass) fired plants	Energy transmission sector
9	Local renewable resources for local demand	Enhancing local renewable use, such as solar, wind, biomass, and others	
10	Next-generation fuels	Development of a carbon-free hydrogen and/or biomass-based energy supply system with required infrastructure	
11	Labeling to encourage smart and rational choices	Publicizing of energy use and CO_2 costs information for smart choices of low-carbon goods and services by consumers and public acknowledgement of such consumers	All sectors
12	Low-carbon society leadership	Human resource development for building a "low-carbon society" and recognizing extraordinary contributions	

for photothermal and photovoltaic use, which is estimated to offer about 1.567 million GWH per year (Huang et al. 2008); (2) wind energy, which, if fully utilized, could lead to a power generation of 3000 GWH or more per year (Yu 2006); (3) biomass energy, which could produce approximately 253 GWH per year if half of all agricultural residues is used by a biomass power plant as fuel (Huang et al. 2008); and (4) tidal and wave energy, which could be equal to 6000 GWH per year (Yu 2006). Huang et al. (2008) suggested that wind energy and biomass energy be developed first because of their relatively low cost, with solar energy and tidal and wave energy gradually enhanced in the future.

Finally, it is important that the awareness of the local population is improved. During our field survey in the summer of 2014, it was discovered that consciousness of environmental conservation is very weak and that locals do not have sufficient sense of urgency for pursuing a "low-carbon society." Without understanding and support from the general public, the aforementioned technological and institutional evolution is difficult to put in place. Government officials and relevant decision-makers must keep in mind that environmental education plays an indispensable role if the long-term success of a "low-carbon society" is expected. The promotion of "low-carbon" education by the mass media, as well as through regular community campaigns with appropriate reward/incentive mechanisms, could be an effective approach.

10.5 Conclusions

As the world's largest alluvial island, China's third largest island, and an important part of the Shanghai metropolitan region, Chongming Island possesses valuable ecological functions. However, the inherent problems, as well as the conflict between social and economic development and nature conservation, must be considered in detail. In this chapter, we discuss ways to achieve the sustainable development of Chongming Island from the perspectives of ecological planning, land use/cover change, water management, and the development of a low-carbon society. In terms of ecological planning, a three-stage plan, together with a division of the island into six ecological functional areas and one natural reservation district, provides a practical approach. Land use/cover change in Chongming reflects the effectiveness of the implementation of eco-island planning on the entire island's landscape layout and the protection of ecological features such as cropland and water. However, by following the approach to expanding the urbanized area that is common in most parts of China, encroachment onto cropland and an extensive sprawl of built-up land were found in Chongming. As for water management, the introduction and upgrading of treatment facilities and systems are essential. Additionally, new technologies that allow rainwater use, energy saving, and water reclamation and recycling should be taken into consideration, together with wastewater condition, running costs, and energy consumption. Furthermore, water utilization for the agricultural sector must be further studied and optimized so as to avoid agricultural nonpoint source pollution. With regard to a low-carbon society, actions proposed by the Ministry of the Environment, Japan, could be adopted by Chongming Island, with consideration of specific local circumstances. Note that if long-term success in low-carbon society practice is expected, environmental education is of vital importance.

Acknowledgments This research was supported by the National Natural Science Foundation of China (41401638), China Ministry of Education's Humanities and Social Sciences Project (14YJAZH028), Shanghai Pujiang Program (14PJ1402800), Shanghai Philosophy and Social Sciences Planning Project (2014BCK001), and Innovation Program of Shanghai Municipal Education Commission (14ZS053).

References

Anderson JR, Hardy EE, Roach JT, Witmer RE (1976) A land use and land cover classification system for use with remote sensor data; U.S. Geological Survey Professional Paper No. 964, USGPO, Washington, DC

Huang B, Ouyang Z, Zheng H, Zhang H, Wang X (2008) Construction of an eco-island: a case study of Chongming Island, China. Ocean Coast Manag 51:575–588

IPCC (Intergovernmental Panel on Climate Change) (2013) Climate change 2013: the physical science basis. Cambridge University Press, Cambridge

Ministry of the Environment Japan (2007) Building a low carbon society: first draft. Available at: https://www.env.go.jp/earth/info/pc071211/en.pdf

Ni X, Wu Y, Wu J, Lu J, Wilson PC (2012) Scenario analysis for sustainable development of Chongming Island: water resources sustainability. Sci Total Environ 439:129–135

Normile D (2008) China's living laboratory in urbanization. Science 319(5864):740–743

NIES (National Institute for Environmental Studies), Kyoto University, Ritsumeikan University and Mizuho Information and Research Institute (2008) Japan Scenarios and Actions towards Low-carbon Societies (LCSs). Available at: http://2050.nies.go.jp/press/070215/file/2050_LCS_Scenarios_Actions_English_080715.pdf

Wang K, Zou C, Kong Z, Wang T, Chen X (2005) Ecological carrying capacity and Chongming Island's ecological construction. Chin J Appl Ecol 16(12):2447–2453

Yu S (2006) Chongming Island: a future base of new and renewable energy in China. Energy Technol 27(1):13–19

Yuan W, James P, Hodgson K, Hutchinson SM, Shi C (2003) Development of sustainability indicators by communities in China: a case study of Chongming County, Shanghai. J Environ Manag 68:253–261

Zhao B, Kreuter U, Li B, Ma Z, Chen J, Nakagoshi N (2004) An ecosystem service value assessment of land-use change on Chongming Island, China. Land Use Policy 21:139–148

Chapter 11
Proposal of a New Water Recycling System Featuring "Water Reclamation and Reuse"

Jun Nakajima and Toshiyuki Shimizu

Abstract Recently, the development of a stable water supply that responds to increasing water demand has been difficult because the spatial-temporal unfair distribution of water resources in the world has expanded due to climate change. Therefore, in regions with a water shortage, it is becoming more and more important to reclaim and reuse water. In this chapter, we first introduce our interdisciplinary research project that has been focused on water reclamation/reuse. The project consisted of four groups: "Reclamation and Recycle Group," "Material Recycle Group," "Green Space Creation Group," and "Management and Policy Group." Then, we introduce the concept of greywater for water reclamation/reuse and show the characteristics of greywater quality in general households of semi-urban regions in Thailand, experimental results from laboratory scale experiments, and acceptance of treated greywater from residents in three Asian countries. Additionally, we show the current situation of rainwater and reclaimed water uses in university campuses and propose a new strategy regarding eco-houses of the future.

Keywords Water reuse • Water reclamation • Greywater • Rainwater • Green campus • Green building

11.1 Introduction

The spatial-temporal unfair distribution of limited water resources in the world is expanding because of climate change. Therefore, the development of stable water supplies that are capable of responding to increasing water demands becomes more and more difficult. The reclamation and reuse of water is a possible solution to this issue. Reclaimed water can potentially be used for various purposes that do not

J. Nakajima (✉)
College of Science and Engineering, Ritsumeikan University,
Kusatsu, Shiga 525-5877, Japan
e-mail: jnt07778@se.ritsumei.ac.jp

T. Shimizu
Ritsumeikan Global Innovation Research Organization, Ritsumeikan University,
Kusatsu, Shiga 525-5877, Japan

K. Nakagami et al. (eds.), *Sustainable Water Management*,
DOI 10.1007/978-981-10-1204-4_11

require potable (i.e., drinkable) water; however, additional studies are required to investigate the applicability of reclaimed water. Greywater, which refers to all wastewater generated by buildings with the exception of toilet wastewater, is thought to be suitable for reuse because it usually contains low concentrations of pathogens. Recently, various measures for energy and water savings have been implemented to reduce the carbon footprint of houses and other buildings. Going forward, the increasing promotion of eco-houses using the thermal capacity of water and green buildings that incorporate roof greening to reduce the electricity consumption of indoor air conditioning and use potable water for environmental purposes in housing estates is expected. However, the development of a method for the appropriate drawing, disposal, and management of sludge generated by water treatment is the most important barrier to the implementation of sustainable water treatment and reclamation.

In this chapter, we first introduce our interdisciplinary research project that has been conducted with a focus on water reclamation/reuse. Then, we discuss the quality of greywater, report the experimental results of greywater treatment studies, and discuss the acceptance of treated greywater from residents. Additionally, we discuss the current situation of rainwater and reclaimed water uses at university campuses and describe new strategies regarding eco-houses of the future.

11.2 Research Project "Water Reclamation and Recycling in Asia"

11.2.1 Background of the Project

Water shortage is a common issue in the arid regions of countries in the Asia Pacific region, which include both developing and developed/emerging countries. In particular, the water shortage issue tends to be problematic in the semi-urban areas of developing countries, since it is not possible to develop large-scale infrastructure systems because of the large investment and capital required. However, some developed countries such as Japan also have issues because of inevitable decreases in infrastructure budgets, which may cause readjustments from a large-scale infrastructure system to a decentralized system in order to reduce operation and maintenance costs. Also, it is important to use limited water resource efficiently for introducing decentralized system successfully. For these reasons, the development of a water circulation system including a new water supply and reclamation system is required on a global scale.

Water reclamation systems are expected to improve the water shortage issues in households; however, the acceptance of the installation of such water reuse systems depends on not only the water supply/demand condition and needs of water reclamation and circulation but also the histories and culture of the residents, as there have been admissibility conditions from users regarding the acceptance of using treated wastewater including blackwater (toilet wastewater) and greywater (all other household wastewater). The major use of treated wastewater worldwide is for agricultural irrigation. However, in Japan the intended uses have been limited to flushing toilets in buildings and for other environmental uses such as pond and creek in the park. The reason for this is that the water used for agriculture is typically guaranteed by water right in Japan.

The water reclamation innovations using membranes have been developed. However the actual conditions of water uses and wastewater discharges are diverse in the Asia Pacific region. Therefore, investigation of the various applications of water that is reclaimed and reused will be required. These investigations will assist with studies regarding the various inflection methods of reclaimed water to develop eco-houses, green buildings, and a green society.

Regarding the other issues in water treatment systems, the pollution of groundwater and soil in developing countries is frequently caused by the inappropriate disposal of drawn waste sludge. In one case, collected sludge was dumped directly on the ground and subsequently infiltrated to the soil, even though a vacuum truck introduced by the Japanese Official Development Assistance (ODA) office was used. Therefore, resource reclamation circulation systems should be researched and developed to meet the economic base in different countries and regions.

11.2.2 Objectives of the Project

As mentioned above, in order to mitigate the serious water issues accelerated by climate change, new water cycles should be established by water reclamation methods and the reuse of used water. As shown in Fig. 11.1, this project researched and developed an innovative water reclamation technology and water recycling system and investigated their application with regional water management policies. The project also developed designs for green building facilities and recycling systems for the waste sludge produced in the water reclamation process. The integrated package proposals should contribute to the establishment of water infrastructure industries and integrated water businesses in the future in Japan. In this project, the objectives were as follows:

1. Development of an innovative water reclamation technology applicable to domestic water usage in Asia and the Pacific regions
2. Design of green building and landscape facilities using reclaimed water that was acceptable by regional people
3. Establishment of a sustainable material recycling system to receive and process sludge waste from the water reclamation process
4. Proposal of policies enabling the water reclamation and recycling system to be sustainably used by the people and society in the region
5. Preparation of a package scenario of water reclamation and recycling

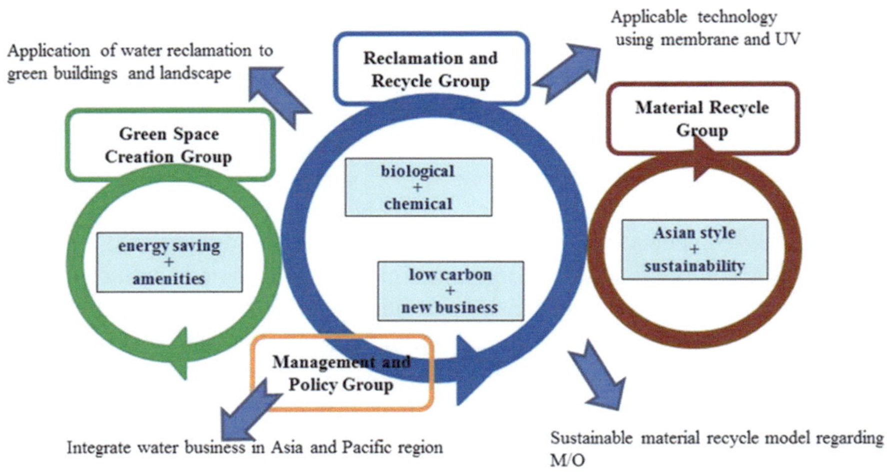

Fig. 11.1 Outline of the research project

Four research groups were established to achieve the project objectives. Each group promotes their own research and personnel interchange between the groups for research revitalization.

11.3 Reclamation and Reuse of Greywater

11.3.1 Why Greywater?

In general, domestic wastewater is categorized as either blackwater or greywater. Blackwater is wastewater from toilets, and greywater is wastewater from all other parts of the household, including the kitchen, bath, shower, and laundry areas. Greywater usually contains low concentrations of pathogens, and thus, it is much more suitable for recycling options in households and is more socially acceptable (Nghiem et al. 2006). Combined wastewater involving blackwater and greywater is usually treated in developed countries. However, only blackwater is treated in the rural areas of developing countries, and it may sometimes be discharged without treatment. Although greywater is considered to be a better target for reuse than wastewater containing blackwater, there have been a few applications of greywater treatment. Various decentralized biological treatment systems that have been used to treat domestic wastewater, such as septic tanks, anaerobic baffled reactors, constructed wetlands, waste stabilization ponds, package tanks with aeration (e.g., Johkasou), and membrane bioreactors, are also candidates for greywater reclamation and reuse (Nakajima et al. 1999; Jefferson et al. 2000; Tilley et al. 2008). Blackwater and greywater segregation has been recommended during the biological treatment in order to reach the ideal quality of the final effluent for reuse purposes

Table 11.1 Characterization of greywater quality

	Categories of greywater						
	Results of individual houses						Results of apartment
Parameter	Cooking	Cooking containing rice washing water[a]	Dishwashing	Bathing	Laundry washing	Laundry washing containing poly-P>3[b]	Combined greywater
pH	6.7±0.8	6.2±0.9	6.7±0.7	7.5±0.4	8.4±0.9	–	7.2±0.4
BOD	360±370	590±410	490±550	160±97	290±240	–	90±44
COD	580±960	1100±1400	990±1500	270±150	800±650	–	160±64
MBAS	1.3±2.5	0.9±0.6	170±240	4.3±5.0	150±120	–	5.0±4.8
TP	7.2±10.2	15±13	2.2±3.8	0.2±0.4	6.1±10	20±16	1.5±0.9
Poly-P	3.6±5.4	7.5±6.7	0.4±1.2	0.0±0.0	3.4±8.5	15±14	0.3±0.5
SS	380±600	410±310	210±260	100±63	250±210	–	69±31
n-Hex	85±260	58±37	200±480	–	–	–	–

Values include the averaged SD in units of mg/L, with the exception of pH
$n=40$ for individual houses except for the BOD analysis ($n=26$–34)
$n=20$ for apartments
[a]Wastewater containing rice washing water within the cooking wastewater [$n=15$ except BOD ($n=12$)]
[b]Wastewater containing poly-P>3 mg/L within the laundry washing wastewater ($n=8$)
BOD biochemical oxygen demand, *COD* chemical oxygen demand, *MBAS* methylene blue active substances, *n-Hex* n-hexane extractable material, *Poly-P* polyphosphate, *SD* standard deviation, *SS* suspended solids, *TP* total phosphorus

(Mah et al. 2009). Additional studies on the feasibility and design of biological treatment technologies for greywater reclamation and reuse will be necessary because the quality of greywater and acceptance of treated greywater are different in each region. Also, the water use situations and greywater characteristics and its quantity in the area should be determined to develop appropriate systems and technologies.

11.3.2 Water Quality of Greywater

11.3.2.1 Raw Water Quality in Domestic Houses in Thailand

Greywater collected from cooking, dishwashing, bathing, laundry washing (four categories) in the individual households, and combined greywater collected from the apartment buildings displayed highly variable parameters (Table 11.1) (Jiawkok and Nakajima 2012). The pH values were almost neutral for the combined greywater and cooking, dishwashing, and bathing wastewater and weakly alkaline for the laundry wastewater. The biological oxygen demand (BOD), chemical oxygen demand (COD), and suspended solids (SS) were high in the cooking, dishwashing,

Table 11.2 Removal ratio of MBAS and TOC in laboratory scale experiments

Parameter	Influent (mg/L)	Effluent (mg/L)	Removal ratio (%)
MBAS	33.9±3.9	3.1±0.7	90.9
TOC	27.7±2.9	10.3±2.4	62.9

Under oxic conditions, hydraulic retention time (HRT): 1 day

and laundry wastewater and lower in the bathing wastewater. The BOD/COD ratio was almost larger than 0.5, with the exception of the laundry wastewater (0.36). This suggested that the laundry wastewater contained a high percentage of sparing or nonbiodegradable organics. The methylene blue active substance (MBAS) concentration was high in the laundry and dishwashing wastewater and low in the cooking and bathing wastewater. The high pH and low biodegradability of the laundry wastewater were attributed to the use of laundry detergents containing surfactants.

The total phosphorus (TP) and polyphosphate (poly-P) concentrations were high in the cooking and laundry wastewater and low in the bathing wastewater. Some detergents used in the area contained sodium tripolyphosphate as a builder to reduce the hardness of the water. The high TP concentration caused by the high poly-P in the laundry wastewater showed the influence of this builder compound. On the other hand, the high concentrations of TP and poly-P in the cooking wastewater were likely caused by food processing wastes, and the cooking wastewater containing rice washing water showed extremely high concentrations of TP and poly-P.

The n-hexane extractable (n-Hex) material concentration was high in the cooking and dishwashing wastewater, suggesting that they contained oil and grease. The combined greywater collected from the apartment buildings was less polluted with a BOD of 90 mg/L compared to the four categories of greywater tested.

11.3.2.2 Treatment of Greywater in Laboratory-Scale Experiments

In this section, the results of the laboratory-scale experiments to test the treatability of greywater using a biofilm process with aeration are shown. The hydraulic retention time (HRT) was 1 day. The dissolved oxygen (DO) concentration in the reactor remained near 8 mg/L. The synthetic greywater for these laboratory scale experiments was prepared by mixing three types of detergents (liquid shampoo, liquid dishwashing detergent, and powdered laundry detergent) into tap water in consideration of the various daily activities associated with the modes of bathing, dishwashing, and laundry. The selected detergents used in this study were commercial detergents that have been widely used in households in Thailand.

The removal efficiencies of MBAS and total organic carbon (TOC) under oxic conditions are shown in Table 11.2. The MBAS removal was greater than 90 %, and the TOC removal was greater than 60 %, indicating that the removal efficiency for

TOC was rather low compared to the removal of MBAS. This result suggested that a contact aeration biofilm system could effectively decrease the toxicity caused by detergents in greywater.

11.3.3 Acceptance of Reclaimed Water in Asian Countries

Figure 11.2 shows the differences in acceptance of reclaimed water use between Thailand, Lao PDR (People's Democratic Republic), and Indonesia. Figure 11.2a shows the greywater reuse acceptance, and Fig. 11.2b shows the statistics of blackwater reuse acceptance after proper treatment. The figure emphasizes that there might be a water shortage problem in these Asian countries. In all three countries, the people's acceptance of greywater for reuse was higher than that of blackwater. In Indonesia, the acceptance of using treated blackwater was quite low, and the acceptance of greywater was lower than the other two countries. A higher percentage of people generally accepted the use of treated water for non-potable purposes such as plant watering, toilet flushing, and car washing.

Treated wastewater is widely used for toilet flushing in many countries such as Australia, Canada, England, France, Germany, Japan, and the USA (Asano et al. 1996; Lazarova et al. 2003). Toilet flushing with greywater is socially acceptable in many countries. However, the acceptance for toilet flushing with reclaimed water was not high in the three countries listed above. In Thailand, more than 80 % of households accepted using treated greywater, while approximately 50 % accepted using treated blackwater containing night soil. Treated greywater was generally chosen for plant watering (approximately 70 %) and floor and car washing (50–60 %). Respondents were more willing to use greywater for daily purposes in which the water did not directly come into contact with the human body (Jiawkok et al. 2013). In Lao PDR, more than 65 % of households accepted using treated greywater, while approximately 50 % accepted using treated blackwater. The acceptance of greywater for plant watering was approximately 50 %; toilet flushing, floor washing, and car washing were 30 %; and other purposes were less than 10 %. In Indonesia, approximately 50 % of households accepted using treated greywater, while more than 80 % did not accept using treated blackwater. The acceptances of using treated blackwater were less than 10 % in all purposes. The acceptances of greywater for plant watering and toilet flushing were approximately 30 %; toilet flushing, floor washing, and dishwashing were 20 %; and other purposes were less than 10 %.

The results obtained in the study and the respondents' acceptance of greywater suggested that an increase in water reuse in these areas would be possible through the introduction of a suitable reclamation system. Increasing the awareness of proper wastewater management is needed in order to motivate people to reuse their household wastewater to save water and reduce the release of wastewater that will negatively affect the environment.

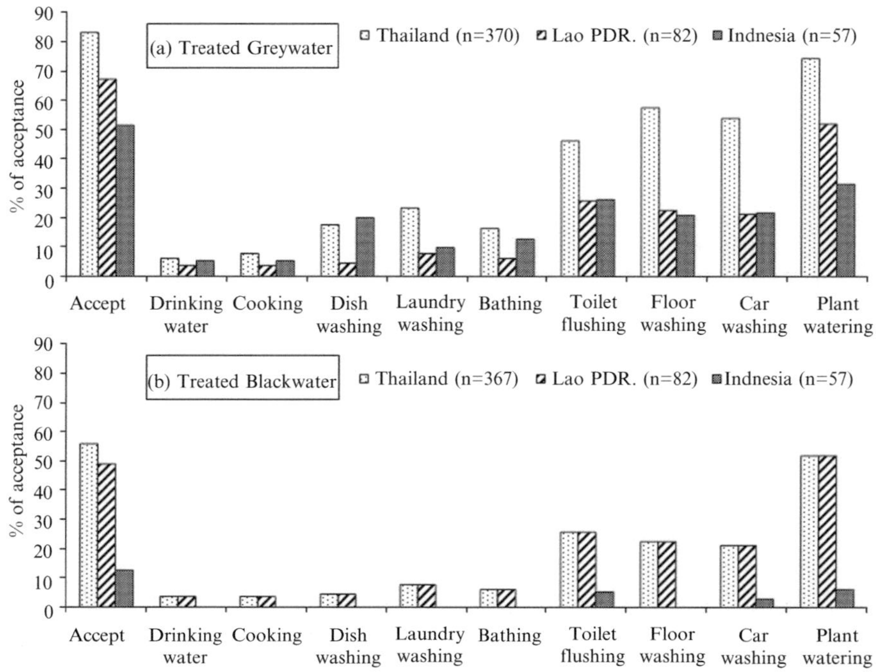

Fig. 11.2 Acceptance of reclaimed water use in three Asian countries

11.4 Research Project "Water Reuse and Rainwater Use in University Campuses for Development of Green Campus"

11.4.1 Introduction

Recently, the importance of environmental aspects in the construction, development, and management of sustainable/green university campuses has been strongly recognized. As part of an action to assist with the development of a green campus, measures to reduce water consumption have been conducted at many universities. There are two methods to reduce water consumption. The first is installing water-saving fixtures such as toilets, urinals, lavatory faucets, and showerheads, and the second is using rainwater or reclaimed water. In recent years, some university buildings have received the certification of green buildings such as LEED (Leadership in Energy and Environmental Design; US Green Building Council) and Green Star (Green Building Council of Australia). "Water efficiency" is one of the evaluation criteria for the certification of green buildings, encouraging the installation of water-saving equipment and systems of using rainwater/reclaimed water. University campuses use large amounts of water and also generate large amounts of waste blackwater and greywater.

11.4.2 Raw Water and Intended Use of Reused Water in University Campuses

Many university campuses have installed systems for using rainwater/reclaimed water, but there is no knowledge regarding the number of systems that are installed and how they are implemented at university campuses. The websites of universities were surveyed in order to obtain knowledge of the approximate number of universities using rainwater or reclaimed water systems and information regarding the intended use of the systems (Shimizu et al. 2015).

In this section, information from universities located in Japan, the UK, Australia, the USA, and Canada are shown. Technical terms regarding the kind of raw water and intended use differed slightly according to the country and university. In this study, the terms were defined as shown in Table 11.3, and water for reuse excluding rainwater was collectively defined as "reclaimed water."

Table 11.4 shows the summary of rainwater and reclaimed water use situations regarding raw water and the intended uses. At Japanese universities, most of the

Table 11.3 Definition of terms

	Category	Remark
Raw water for reuse	Rainwater (RW)	Roof water and storm water
	Greywater (GW)	Wastewater from kitchens, showers, etc.
		(Treated at university campuses)
	Wastewater (WW)	Blackwater and greywater
		(Treated at university campuses)
	Air conditioner (AC)	Used water from air conditioners
		(Treated at university campuses)
	Recycled water (ReW)	Treated water from wastewater treatment plants located outside of the university campus
	Others (O)	Experimental wastewater, industrial water, etc.
Intended use of rainwater/reclaimed water	Flushing toilets (FT)	Water for flushing toilets
	Landscape irrigation (LI)	Watering/irrigation for plants, gardens, etc.
	Environmental use (EU)	Fountains, ponds, creeks, waterfalls, etc.
	Air conditioning (AC)	Cooling towers, roofs, etc.
	Washing outdoor (WO)	Washing vehicles, walls, canoes, etc.
	Others (O)	Fire-fighting, washing experimental instruments, etc.

Table 11.4 Summary of rainwater and reclaimed water use situations regarding raw water and intended uses at university campuses

		Japan		UK		Australia		USA and Canada	
Countries		*n*	(%)	*n*	(%)	*n*	(%)	*n*	(%)
Raw water	Rainwater	107	(92.2)	15	(100)	29	(100)	29	(69.0)
	Greywater	8	(6.9)	–	–	5	(17.2)	5	(11.9)
	Wastewater	4	(3.4)	–	–	1	(3.4)	2	(4.8)
	Air conditioner	8	(6.9)	–	–	3	(10.3)	13	(31.0)
	Recycled water	4	(3.4)	–	–	4	(13.8)	16	(38.1)
	Others	13	(11.2)	–	–	–	–	1	(2.4)
Intended use	Flushing toilets	95	(81.9)	15	(100)	22	(75.9)	13	(31.0)
	Landscape irrigation	60	(51.7)	–	–	25	(86.2)	35	(83.3)
	Environmental use	6	(5.2)	–	–	5	(17.2)	3	(7.1)
	Air conditioning	9	(7.8)	1	(6.7)	5	(17.2)	14	(33.3)
	Washing outdoor	–	–	–	–	4	(13.8)	4	(9.5)
	Others	6	(5.2)	–	–	2	(6.9)	2	(4.8)
Total number of samples		116		15		29		42	

water reclamation systems currently use rainwater rather than other types of reclaimed water. The percentage of facilities that have introduced treatment systems for greywater or wastewater was less than 10 %. Regarding the intended use, toilet flushing was the largest use (~82 %), and ~52 % of the use was for landscape irrigation. As for other intended purposes, rainwater or reclaimed water was used for environmental uses and air conditioning. In the UK, all universities used rainwater only, and almost all of the universities used rainwater or reclaimed water for flushing toilets only. In Australia, rainwater was the largest resource of raw water, and all universities that were using rainwater or reclaimed water used rainwater. The percentage of greywater use was relatively higher than the other countries surveyed. Regarding the intended use, landscape irrigation was the largest use of water, and flushing toilets was the second highest use. In the USA and Canada, the percentage of universities using rainwater was relatively lower than the other countries, even though rainwater was the largest resource. As a characteristic trend, the percentages of campuses using air conditioners and recycled water were higher than in the other countries. Regarding the intended uses, landscape irrigation was the largest use, and flushing toilet use was lower than in the other countries.

In Australia, the USA, and Canada, many universities used rainwater or reclaimed water in combination with other kinds of raw water, and the intended uses were also used in combination. Figure 11.3 shows the major combinations in each country.

The results discussed above showed that the type of raw water and its intended use on rainwater/reclaimed water use at university campuses varied according to the country. The status of water resources such as rainfall was considered to affect the diversity of rainwater/reclaimed water use.

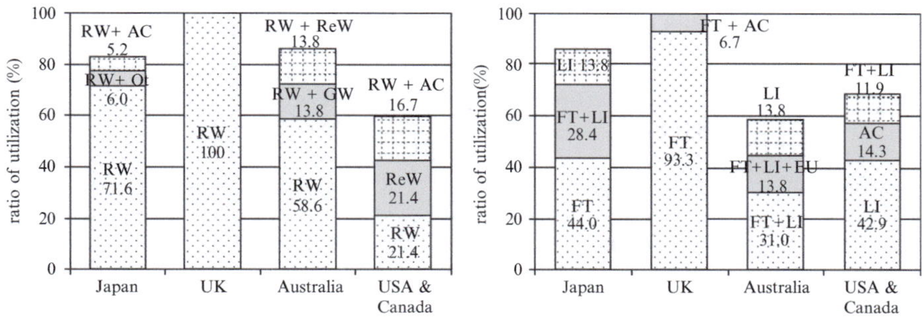

Fig. 11.3 Major combinations of raw water and intended uses (*left*: raw water, *right*: intended use)

Fig. 11.4 *Left*: Environmental pond for rainwater storage (no water on fine days). *Center*: Plant irrigation wall. *Right*: Jokaso (treatment facility for greywater from the building)

11.4.3 Green Building at a University Campus: Case Study at Ritsumeikan University

The new research building for the departments of Civil Engineering, Environmental Systems Engineering, and Architecture and Urban Design was constructed at the Biwako-Kusatsu Campus, Ritsumeikan University, in 2014. The building was designed as an experimental building in which new technologies for sustainable campuses could be innovated and tested. By introducing the latest technologies, including building equipment and construction materials for energy savings and the reduction of environmental loads, students and teachers can advance research for their results, verification, and improvement. In this building, students and teachers are researchers and subjects.

The technologies are supported by the collaboration between Ritsumeikan University and 20 companies. The results obtained here will be open to the world. The building is using rainwater for environmental use (environmental pond) and plant irrigation on a wall, and a jokaso is installed for reclamation/reuse experiments of greywater (Fig. 11.4). The building received the Award of Low Carbon Building from the Shiga prefectural government in 2015.

11.5 Eco-Houses of the Future: Enemane House Project

11.5.1 What Is the Enemane House Project?

Because more than 30 % of the total energy in Japan is consumed by households and the buildings sector (ANRE 2015), the reduction of energy consumption in housing is important. Public attention regarding energy security has been increasing, and the attention to the "net zero energy house (ZEH)," which performs local production for the local consumption of energy, is rising in the housing sector.

The Enemane House project (SII 2015) is a competition of the construction of a ZEH among five universities in Japan and has been carried out from 2014 by the Agency for Natural Resources and Energy, Japan government. The theme of the project is "Future House Considered by Students," and the purpose is to build and display five model houses that propose advanced technologies and new lifestyles under the concepts of "energy," "life," and "Asia," as described below. In 2015, the five houses were built in Yokohama City.

1. Energy:

   To achieve daily life activities with 20–30 % energy consumption compared to normal houses by installing the advanced technologies and incorporating the passive design

2. Life:

   To create a new premium for houses and communication by the construction of lifestyle and service innovations for realization of high-quality life and resilient social systems that support it

3. Asia:

   To achieve the transfer of the concept, parts, and a part of the system or whole by the technology development and proof focused on expansion to Southeast Asia

11.5.2 Concept of the Ritsumeikan Team

Although the energy-saving measures in the houses have attracted attention, water reclamation and reuse in the residential system have not been as well discussed. Therefore, the Ritsumeikan team proposed the concept of "ZEH + water." Figure 11.5 shows the concept, and Fig. 11.6 shows the model house of the Ritsumeikan team. A box-in-box structure, which is an effective technique for improving the indoor thermal environment, was selected as the basic space arrangement in the construction of the ZEH, and energy-saving technologies such as heat insulating methods/fittings and a transpiration louver using the Shigaraki ware pottery that has been studied at Ritsumeikan University were installed. In addition, a photovoltaic

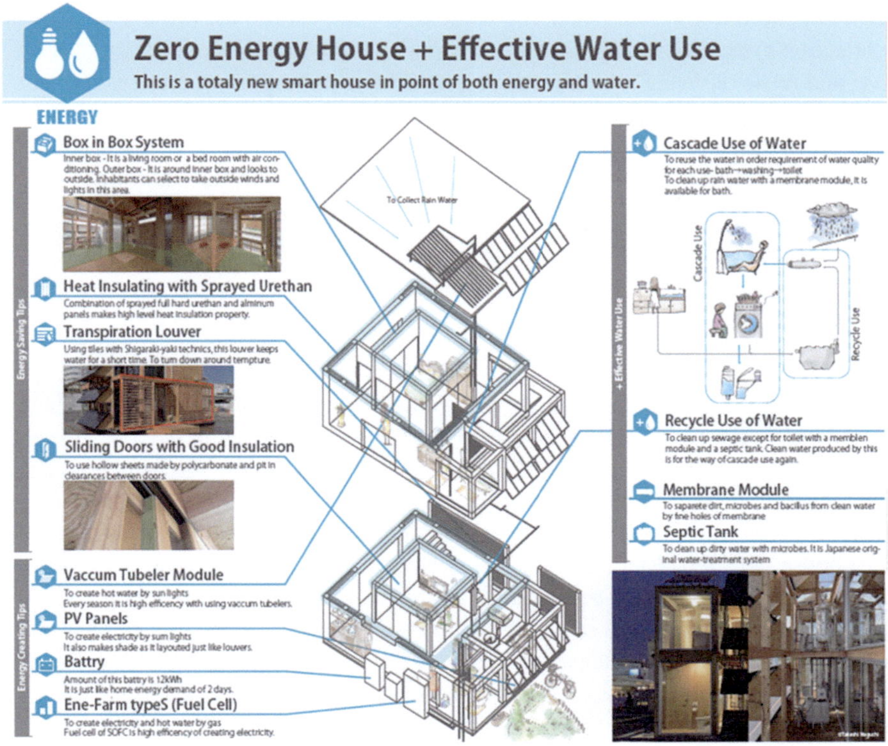

Fig. 11.5 Outline of the Ritsumeikan team in the Enemane House project

Fig. 11.6 Model house built by the Ritsumeikan team at Yokohama City

power generation system, fuel cell system, rechargeable battery, and solar heat collection system were also installed. For saving the water consumption, a "cascade use and reuse of water" system was proposed by installing water reclamation technologies including a membrane module and biological treatment system. Rainwater and reclamation/reuse systems in the houses were thought to be a self-sustaining infrastructure. Namely, our proposal aspired to develop highly self-sustaining housing from the viewpoints of energy and water.

11.6 Summary

The application of water reclamation and reuse systems will be successful when the systems are accepted by the people who use the reclaimed water. Greywater reclamation and reuse were focused on in this chapter because of the higher potential for acceptability by the people. The possible usage of reclaimed water also depends on people's interest and tolerance. Two examples, a green university campus and an eco-house with zero energy, were studied as applications of greywater reclamation and reuse. Although these potential usages have not obtained enough popularity in Japan, they can be applied in Asian countries in the near future. Appropriate technologies with suitable management systems must be developed to establish sustainable water reclamation and reuse systems in Asia.

References

Agency for Natural Resources and Energy, Japan Government (2015) Energy White Paper 2015. http://www.enecho.meti.go.jp/about/whitepaper/. Accessed in 29 Dec 2015 (in Japanese)

Asano T, Maeda M, Takaki M (1996) Wastewater reclamation and reuse in Japan: overview and implementation example. Water Sci Technol 34(11):219–226

Green Building Council of Australia (website). http://www.gbca.org.au/green-star/. Accessed in 29 Dec 2015

Jefferson B, Laine A, Parsons S, Stephenson T, Judd S (2000) Technologies for domestic wastewater recycling. Urban Water 1(4):285–292

Jiawkok S, Nakajima J (2012) Wastewater treatment using a biofilm process for greywater reclamation and reuse. J Water Treat Biol 48(3):109–116

Jiawkok S, Ittisupornrat S, Charudacha C, Nakajima J (2013) The potential for decentralized reclamation and reuse of household greywater in peri-urban areas of Bangkok. Water Environ J 27:229–237. doi:10.1111/j.1747-6593.2012.00355.x

Lazarova V, Hills S, Birks R (2003) Using recycled water for non-potable, urban uses: a review with particular reference to toilet flushing. Water Sci Technol 3(4):69–77

Mah DYS, Bong CHJ, Putuhena FJ, Said S (2009) A conceptual modeling of ecological greywater recycling system in Kuching city, Sarawak, Malaysia. Resour Conserv Recycl 53(3):113–121

Nakajima J, Fujimura Y, Inamori Y (1999) Performance evaluation of on-site treatment facilities for wastewater from households, hotel and restaurants. Water Sci Technol 39(8):85–92

Nghiem LD, Oschmann N, Schafer AI (2006) Fouling in greywater recycling by direct ultrafiltration. Desalination 187:283–290

Shimizu T, Nakajima J, Chikamoto T, Nakagami T (2015) Rainwater and reclaimed water uses in domestic and foreign university campuses. J Water Waste 57(9):42–50 (in Japanese)

Sustainable Open Innovation Initiative (SII), Website of the Enemane House 2015. https://sii.or.jp/emh2015/. Accessed 10 Jan 2016 (in Japanese)

Tilley E, Lüthi C, Morel A, Zurbrügg C, Schertenleib R (2008) Compendium of sanitation systems and technologies. Swiss Federal Institute of Aquatic Science and Technology (Eawag), Dübendorf

U.S. Green Building Council (website). http://www.usgbc.org/leed. Accessed 29 Dec 2015

Part V
Conclusions

Chapter 12
Conclusions: The Future of Sustainable Water Management

Malcolm Cooper

Abstract Ensuring sufficient unpolluted water for urban, agricultural, and industrial use is arguably the most important issue facing the world's communities. Countries such as Australia are currently unable to supply enough water to many areas in times of drought, and it is estimated that perhaps a billion people in the Asia-Pacific Region face the prospect of unsafe water at all times. This chapter and book take a new and skeptical look at some of the underlying factors that affect the management of this vital resource and the proposed solutions. Traditionally, water management policies and practices have dealt only with problems of water distribution to meet the ever-increasing demand, rather than better management of existing resources. The largely fragmented approach that results has contributed to the overexploitation of water resources. Nevertheless, in many parts of Australia, China, South Africa, Canada, the United States, Europe, Japan, and elsewhere, efforts are currently being made to better manage water distribution systems using proactive methods instead of simply reacting to supply and demand problems. Proactive management methods include new ways of accounting for water and methods for reducing losses, as well as benchmarking against international high performers. However, there are no across-the-board solutions since context matters and managers must therefore learn from local operating experience. The difficulties experienced by integrative efforts in this situation indicate that a significant part of the problem lies in the structures of governance in the water industry. Water management should be a regional, national, and international level concern, and it is in many places but generally at the level of policy rather than responsibility for infrastructure and pricing of water.

Keywords Demand • Integrated water management • Governance • Fragmentation • Policy

---

M. Cooper (✉)
Graduate School of Asia Pacific Studies, Ritsumeikan Asia Pacific University, 1-1 Jumonjibaru, Beppu, Oita 874-8577, Japan
e-mail: cooperm@apu.ac.jp

K. Nakagami et al. (eds.), *Sustainable Water Management*,
DOI 10.1007/978-981-10-1204-4_12

12.1 Introduction

If we want to correctly determine the future of water management, we first need to understand the existing situation (Somlyody 1994; Gleick 2000; Steiner et al. 2000; Duda and El-Ashry 2000; GWP 2003). This understanding can be obtained through a series of propositions (this is a partial list only, intended to be instructive in the debate regarding the future of water management).

12.1.1 Propositions

- Fresh water is arguably the most important natural resource on Earth, yet it is currently squandered in vast quantities.
- While demand outpaces supply in some parts of the world, in almost all places, water is wasted in inappropriate applications and as a result of inadequate/non-existent/deteriorating infrastructure, especially drinking-quality water.
- The current situation is compounded by the pollution levels evident in many water supply systems and the increasing impact of climate change.
- As the volume and cleanliness of available water supplies decline in many countries, governments must effectively deal with this challenge. Reliable information on availability and use is one of the keys to shaping better policies that provide equal access to and sustainable use of increasingly problematic water resources. However, the need for better policy and governance structures is equally important, if not more so.
- The context of the water management problem matters. Local capacity to ensure effective management is a crucial variable, especially the ability to learn from operating experience (Garrick 2015).

These propositions should be evaluated in the context of a series of scenarios that must be addressed to successfully provide the best conceptual and practical approaches to dealing with the various problems in water demand and supply management.

12.1.2 Scenarios

- The amount of global water per capita is expected to fall to 5100 cubic meters per person by 2025, as the world's population grows from six billion to over eight billion (the Asia-Pacific Region in 1999 had only about 3690 m^3) (Hinrichsen n.d.).
- More than 2.8 billion people in more than 40 countries will face water stress or scarcity conditions by 2025 based on United Nations (UN) population projections (UNEP 2006).
- By 2050, the number of people and countries facing water stress or scarcity will increase, affecting 44 % (four billion) of the projected global population of over nine billion (UNEP 2006).

- Many of the affected people and their communities will be located in the Asia-Pacific Region. Also, despite important recent efforts, 2.4 billion people in this region are still without an acceptable means of sanitation (disposal of waste water), while at least one billion do not have access to clean water now.
- Relating these factors to what appears to be happening on a geophysical level, scientists specializing in climate change issues have attempted to forecast the effect of such changes on global water supplies up to approximately 2080. For example, predictions for Australia and a number of other countries indicate that there will be between 25 and 50 % less surface water in 2080, and a similar situation will exist in many parts of Africa and South Asia. However, water supplies will remain approximately the same in the remainder of Asia (Waldron 2007).
- The resulting complexities mean no permanent solutions may be possible, but there must be a strong effort to develop consistent and reliable diagnostic approaches to the water management problem (Garrick 2015).

If these predictions are accurate, the focus on water management will intensify, especially regarding how to better use the declining resource (Gleick 2000; Steiner et al. 2000; Duda and El Ashry 2000; GWP 2003; Waldron 2007; Garrick 2015). Water demand management rather than concentration on delivery (pipes) must become the foundation of managing water supply systems (Waldron 2007).

12.1.3 The Politics of Water Management

Water scarcity and pollution of existing resources are key issues in many regions. Traditionally, communities have dealt with these problems by focusing mostly on increasing the supply to meet the ever-increasing demand for freshwater from the domestic, agricultural, industrial, and service sectors of the economy. The largely fragmented approach traditionally applied to the management of existing water supplies has promoted conflicts and competition between users (there are predictions of "coming water wars" (Hinrichsen n.d.; Cooper 2013)), neglect of collection and distribution systems, and also contributed to the overexploitation of water resources. Many countries currently face the challenge of overcoming fragmented approaches and designing and implementing integrated water management mechanisms, particularly those that implement projects that transcend national boundaries and those within the major cities.

12.1.4 The Evolution of Water Management Practices

Despite current complications, efforts are being made to better manage water distribution systems using proactive, instead of only reactive, methods to solve problems. These proactive management methods include new ways to account for water and methods for loss reduction such as new technologies in water pressure control systems, leakage detection, hydraulic data reporting systems, and benchmarking against international high performers. However, the difficulties experienced by participants in these integrative efforts indicate that a significant part of the problem lies in the structure of governance in the water industry. Many places clearly recognize that water management should be a regional, national, and international level concern, but generally they do so only at the policy level rather than in their actions and assignment of responsibility for infrastructure and water pricing.

For example, Australia has experienced policy confusion in the recent drought/wet cycle in Queensland, the ongoing Murray-Darling saga, and failed attempts to lessen or remove the stifling hold of local government on water management. These issues indicate that the politics of water management is one of the primary reasons for problems in the industry. Much of the wastage of water in Australia is a result of the continuation of the historical arrangements for management and delivery. These are carried out by state and local governments that have, in most cases, proven incapable of discharging this responsibility as a result of a "run-to-fail" approach to the maintenance of their collection and distribution infrastructure.

The short-term view adopted by most councilors and their staff as well as state government officials is of major concern for a type of infrastructure that is costly, requires long-term planning and financing, and is largely invisible. Councils tend to transfer income into general revenue and neglect water infrastructure for more visible infrastructure such as roads, even if capital-debt ratios and income from the water supply system allow effective management of existing infrastructure. In addition to this, state governments cannot agree on water allocation rights, solutions for common salinity problems, or the need to maintain environmental flows in the country's major river systems.

The solution to these problems related to the local control of water supplies often involves corporatization of water supply function within the local government. This removes the direct control of councilors and senior managers of council in favor of actual water industry expertise. Full privatization of the industry is another solution, but the few times this was done were not totally successful. Integrated water resources management (IWRM) is critical at the state level but still remains elusive in many respects (Garrick 2015).

12.2 Integrated Water Resources Management

The first issue regarding water demand management is the claim that water that occurs naturally almost everywhere should be available free of charge or at low cost regardless of any supply difficulties (Cooper 2004). This misconception gives rise to emotional and often highly politicized arguments when communities debate questions related to water use, charges, and the location of water supply services such as dams and storage reservoirs. Water has been traditionally viewed as a commodity with little economic value in most countries throughout the world. The value of water and the costs involved in obtaining it rarely become an issue, even in times of relative scarcity. Therefore, individual and/or community water property rights have always been upheld over conservation needs, even to the detriment of in-stream environmental flows and diminished aquifer storage capacity in river watersheds. Meanwhile, saline intrusion or land subsidence have rarely been viewed as a reason to restrict usage although they may raise the cost of access.

The definition of water scarcity also remains open to debate. Basic economic theory defines scarcity as a point where the supplies of a commodity are lower than the total demand for its use. This definition conjures images for most of situations where freshwater is not available in sufficient quantity to meet the requirements without (unacceptable) radical changes in usage levels and changes in pricing policy. These changes could be designed to promote substitution of treated wastewater for freshwater in agriculture or industry and increase available potable water supplies, yet they are resisted (Kessler 1997).

The remaining concentrations of groundwater and surface water in many parts of the world compound the tendency to assume all water used by communities must be *potable* and cheap. The abundance of supply in some areas, and the use of potable water in industrial and agricultural processes in general, conjures up a favorable situation for all (Hinrichsen n.d.), and therefore, substitutes for potable water are not sought. But, while it is unlikely that many areas will face critical water shortages in the short to medium term, currently relied-upon water supplies are experiencing severe reductions in quality, as seen in Australia. This is economic scarcity (Cooper 2013). Thus, the recent appearance of a desire to address sustainability issues of access, supply, quality, and pricing, along with extensive research into practical solutions for particular supply and distribution problems, is a welcome indication of the recognition that there is a need to address water management problems in the Asia-Pacific Region (as outlined in this volume).

Water supplies in this situation are limited relative to demand at *the price that users wish to pay* (Kessler 1997). This means that some communities may face water shortages because the costs of developing new supplies cannot be funded. Addressing this problem will require basic changes in how community water supplies are perceived and allocated (Kessler 1997; Garrick 2015). A uniform, effective, and comprehensive legal framework covering ownership, supply, pricing, use of water, and disposal of wastewater is an important key to the effective management of these resources in this situation. The framework can be implemented through the development of policies favoring integrated water resources management (see Chap. 2 of this volume).

12.2.1 Governance

Governance of water resources "refers to the range of political, social, economic, and administrative systems that are in place to regulate the development and management of water resources and provision of water services at different levels of society" (GWP 2003). The following issues are important:

- Reliability of information about water resources.
- Level and equitability of access to water.
- Effectiveness of the enabling environment for sustainable water resources management (Garrick 2015). The considerations here include (1) poor enforcement of regulations, (2) problems in the coordination and integration of government agencies, and (3) as a result, weakly integrated water resources management.
- Poor private sector involvement through lack of incentives and integrative policies.
- Participation problems for many stakeholders. Participation is limited to government entities and international organizations, in other words, upward integration, not downward.
- Poor knowledge and understanding of the values and benefits of water, including water pricing, rights, and management technologies (Waldron 2007; Cooper 2013).

The ability to achieve effective system governance is one of the most important yet most complex challenges of integrated water resources management (Conagua 2006). It requires the involvement of all stakeholders in a way that allows them to identify and implement the best possible solutions to water issues that will benefit the majority. Moreover, policy modification is time-consuming, and dealing with many governments requires an understanding of government dynamics, as well as effective persuasion and negotiation skills. Examples of appropriate local governance include neighborhood cooperatives for infrastructure development, local waste recycling schemes, integrated transport plans developed together with user groups, and regional initiatives of state agencies, industrial groups, and residents to control watershed deforestation. At the international level, governance has been viewed primarily as intergovernmental, but it must now involve various nongovernmental organizations (NGOs) and other participants such as businesses. The influence of the important global mass media also interacts with all of these (Conagua 2006).

12.3 The Future of IWRM

Integrated water resources management (IWRM) is "a process that promotes the coordinated development and management of water, land, and related resources to maximize the resultant economic and social welfare in an equitable manner without compromising the sustainability of vital ecosystems" (GWP-SEATAC 2003). It is a political process and involves mediation of conflicting interests (Kataoka 2002; Cooper 2013). The three pillars of IWRM include management instruments, enabling environments, and an effective institutional framework.

In summary, analysis and understanding of IWRM begin by defining a spatial locus within which the critical components interact with and among each other. The most appropriate future locus of analysis and action will be the watershed or catchment area, because it is a naturally defined and discrete unit of the Earth's surface. More concretely, it is a naturally defined territorial unit surface drainage system and as such is a clearly identifiable ecological unit for interface management between biophysical and human systems (Mitchell 1990).

12.3.1 A Watershed Approach

The concept of integrated water management has been evolving over the past half-century (Duda and El-Ashry 2000). During the 1930s, comprehensive water management projects like that in the Tennessee Valley initiated this evolution (Lundqvist et al. 1985). Mitchell (1990) and the Tennessee Valley Authority (1989) showed that integrated land and water resources management policies are necessary. However, even though many academic publications have addressed this subject, it is difficult to achieve in practice. Although we might see trade-offs among economic sectors, full consideration is often not given to issues such as the needs of environmental flows in a river as distinct from its water load, for example.

Failure to achieve integrated management has been attributed to the strength of government ministries responsible for other sectors that oppose the concept, as well as institutional bottlenecks in implementation (Garrick 2015). Somlyody (1994) showed that community-by-community and sector-by-sector approaches that remain the norm are fast becoming a serious international issue. Experience reveals that many water conflicts stem from the implementation of supply-side projects only and that this contributes significantly to the decline of important ecosystems around the world (Duda and El-Ashry 2000).

Therefore, the largely fragmented approach that has traditionally been applied has allowed conflicts and competition and led to the overexploitation of scarce water resources. Many countries are currently challenged by overcoming fragmented subsector approaches and designing and implementing integrated mechanisms, particularly the implementation of projects that transcend subsectors and discrete areas of local government.

The watershed approach to IWRM refers to the formulation and implementation of courses of action involving, as much as possible of the natural and human resources within the confines of such a unit, taking into account the social, political, economic, and institutional factors operating in order to achieve specific objectives. The concept of water resources management within a water catchment area, with a focus on the integration of land- and water-related issues, has been applied in a range of countries including India, Australia, China, and Japan. In India, for example, national water policy mandates that water resources planning be undertaken for a complete hydrological unit, such as drainage watershed or sub-watershed. In Indonesia, institutions for water resources management have been established for some catchments (watersheds), although these have yet to become fully functional (ESCAP 2000).

The rationale for insisting on a watershed base for future IWRM can be summarized:

- The watershed is a functional unit of water assessment established by physical relationships (Steiner et al. 2000).
- The watershed approach is able to evaluate the biophysical linkages of up- and downstream activities, as part of the hydrological cycle.
- The watershed approach is holistic, which means that the many facets of resource development can be included.
- Water collection and other land-use activities often result in a chain of environmental impacts that can best be examined within this context.
- The watershed approach has a strong economic as well as environmental logic. Many of the externalities involved with agricultural, industrial, and residential management practices are internalized when the watershed is managed as a unit.
- The watershed provides a framework for analyzing the effects of community interactions with the environment. Therefore, the environmental impacts within the watershed can operate as a feedback loop for changes in local social systems and
- The watershed provides a context by which integration relevant to IWRM within the natural and human systems can be better understood. The integration of various sector views and interests relevant to IWRM, thus, becomes more tractable in the watershed context.

Another view of the future of IWRM in the watershed context considers the interaction of resources, population, institutions, and technology. At the core of interactions among the components is governance, as noted above. Governance viewed in this light is the hub that balances and harmonizes the interactions. Therefore, it ensures that the demands of the population are being met without endangering the sustainability of the resources, because technologies, policies, and organizational arrangements are applied and appropriately designed to meet the goals of water resources management.

Watershed organization of water resources can take many forms depending on the size of a watershed, the goals of management, the political requirements of a country, and the existing capabilities of resource management. In Southeast Asia, for example, the river watershed organizations have many different models such as committees, commissions, authorities, tribunals, corporations, foundations, and councils. The future use of such a water resources management framework must focus on understanding the level of participation of different actors, sectors, and organizations, and the manner by which efforts and resources are coordinated and harmonized through the councils and other parts of the governance mechanism. Recent discussions of experiences worldwide have confirmed the need for holistic, transdisciplinary, integrated, and participatory approaches.

12.3.2 The Messages in This Book

Many important points are made in this book about what is required to achieve effective water management in the future, and for this summary, I have chosen a few of the most important. Kato et al. (Chap. 2) note that participatory approaches that feature the broad inclusion of local stakeholders have become a basic requirement in any design for future water resources management. However, actual implementation of an effective form of management plan remains one of today's most difficult challenges. The authors argue that the problem exists because "hard-path" water resources management is the preferred course of action (for fixing infrastructure). However, the future requires a soft-path, adaptive management system that considers all aspects of the integration of resource management (Garrick 2015).

For Setiawan et al. (Chap. 3), environmental assessment is one of the main keys for the design of a local framework of integrated water resources management. They correctly point out that in the future we must deal with land use and climate changes as integral parts of our approach to water resources management and not only concentrate on the technical aspects of water collection, cleaning, and distribution. For them and Budiasa and Hisaaki (Chap. 5), the best way to do this is to involve all stakeholders equally, including local people, in acquiring data, in undertaking strengths, weaknesses, opportunities, and threats (SWOT) analyses, and in making decisions. However, this process will be useful and successful only if it has appropriate financial and legal support.

Somewhat similar messages are given by Budiasa et al. in Chap. 7, Li and Nakajima in Chap. 8, and Han and Qian in Chap. 10. The additional points made are critical. The key issues are not so much about the water, but about the institutional legal water regulatory arrangements, pollution control, and chosen approaches to water resources allocation and management. The implementation of integrated management action plans is as important as seeking new sources of water to achieve the most optimal multifunctionality and effectiveness in water management.

12.4 Conclusions and the Future

Experience demonstrates that technological developments or changes in the infrastructure of the water industry do not succeed without effective policy, institutional, and legal structures to support them (Cooper 2013). Our use of water resources is closely linked with that of all other resources. All must be managed together for biodiversity preservation and community participation area by area, because opportunity costs, social institutions, and environmental requirements are different on a country-by-country basis. The approach of integrated water management was developed to provide a solution to this, but results have been disappointing, since not all economic activities, land use considerations, biodiversity needs, and political debates have become integrated with the water security issue. However, as this discussion has shown, a more comprehensive approach is now developing at both the national and local levels. This approach may achieve a more strategic implementation of policy reforms, better technical programs for integrated water and wastewater management, and the introduction of projects supportive of a collaborative response to actual or potential water crises that is required around the world.

The responsible use of water requires ensuring the amount withdrawn from a source does not exceed the environment's capacity to renew that quantity or the environment's needs itself, that the resource is equitably and efficiently collected and distributed (shared), and that there are dispute mechanisms available to adjudicate transboundary, overuse, and intergenerational use issues (Kessler 1997). Experience also demonstrates that no technological or infrastructure investments to achieve better collection and distribution will succeed without effective local policy, institutional, and legal structures. Also, domestic water policies and actions must go hand in hand with international policies and relationships as globalization proceeds (Duda and El-Ashry 2000). Effectively addressing the aspects of water supply and demand, land use, ecosystem health, and socioeconomic development in a more coherent and linked framework will facilitate the needed transition to a new development paradigm. We need integrated and holistic policies that will allow steady improvement in water quality and availability standards, without destroying the biological diversity on which our economies and lives rest.

References

Conagua (2006) Asia Pacific regional document. 4th World Water Forum. Commission Nacional del Agua, Mexico City

Cooper M (2004) The economics of water loss management, Manual 3. Sustainable Industries Division, Queensland Environmental Protection Agency and Wide Bay Water, Brisbane

Cooper M (2013) Water: looming starvation in the midst of plenty–water management policies and practices in the Asia Pacific Region. Paper presented to ICSS-Asia 2013, ANU, Canberra, Australia, 4–8 February

Duda AE, El-Ashry MT (2000) Addressing the global water and environment crises through integrated approaches to the management of land, water and ecological resources. Water Int 25(1):115–126

Garrick DE (2015) Water allocation in rivers under pressure. Edward Elgar, Cheltenham

Gleick PH (2000) The changing water paradigm: a look at twenty-first century water resources development. Water Int 25(1):127–138

Global Water Partnership (GWP) (2003) Effective water governance: action through artnerships in central and Eastern Europe. Status Report to the 3rd World Water Forum in Kyoto, Japan, 16–23 March

Global Water Partnership Southeast Asia Technical Advisory Committee (GWP SEATAC) (2003) Water for the 21st century: vision to action

Hinrichsen D (n.d.) Freshwater: lifeblood of the planet. People & Planet.net. http://www.peopleandtheplanet.com/index.html@lid=26385&section=38&topic=44.html. Accessed 28 Jan 2015

Kataoka Y (2002) Overview paper on water for sustainable development in Asia and the Pacific. Asia-Pacific Forum for Environment and Development First Substantive Meeting, Bangkok, Thailand, January 12–13 Jan 2002

Kessler P (1997) Economic instruments in water management: a more efficient approach. In: Kay M, Franks T, Smith LED (eds) Water: economics, management and demand. Chapman Hall, London

Lundqvist J et al (1985) Strategies for river watershed management: environmental integration of land and water in a river watershed. Reidel, Lancaster

Mitchell B (1990) Integrated water management: international experiences and perspectives. Bellhaven Press, London

Somlyody L (1994) Water quality management: can we improve integration to face future problems? In: Living with water – an international conference on integrated water resources management. RAI, Amsterdam, pp 341–360

Steiner F, Blair J, McSherry L, Guhathakurta S, Marrufo J, Holm M (2000) A watershed at a watershed: the potential for environmentally sensitive area protection in the upper San Pedro Drainage Basin (Mexico and USA). Landsc Urban Plan 49(3–4):129–148

Tennessee Valley Authority (1989) Integrated regional resources management. TVA, Knoxville

UNEP (2006) Global international waters assessment. UNEP, Nairobi

UNESCAP (2000) State of the environment in Asia and the Pacific Region 2000. UNESCAP, Bangkok

Waldron T (2007) Water demand management and associated environmental impacts. Presented to the 1st Asia-Pacific Water Summit, Beppu, Japan, 3–4 December

Index

K. Nakagami et al. (eds.), *Sustainable Water Management*,
DOI 10.1007/978-981-10-1204-4

MIX
Papier aus verantwortungsvollen Quellen
Paper from responsible sources
FSC® C105338

If you have any concerns about our products,
you can contact us on
ProductSafety@springernature.com

In case Publisher is established outside the EU,
the EU authorized representative is:
Springer Nature Customer Service Center GmbH
Europaplatz 3, 69115 Heidelberg, Germany

Printed by Libri Plureos GmbH
in Hamburg, Germany